The Princeton Review

Math Smart
Junior II

The Princeton Review

Math Smart
Junior II

More Math Made Easy

by

Paul Foglino

Random House, Inc., New York 1998

ISBN 0-679-78377-6

Editor: Rachel Warren
Production Editor: Amy Bryant
Production Coordinator: Matthew Reilly

Manufactured in the United States of America.

9 8 7 6 5 4 3 2 1

First Edition

ACKNOWLEDGMENTS

I'd like to thank Evan Schnittman for entrusting me with this project, and my editor, Rachel Warren, whose work made this a better book than the one I wrote. Thanks to Libby, Jenny, and Caroline for their encouragement and suggestions. I'd also like to thank the Princeton Review's editorial and production crew.

CONTENTS

Acknowledgments v

Introduction: Getting to Know Les Ismore ix

Chapter 1 History:
Where Do Numbers Come From? 1

Chapter 2 Words and Numbers:
They're the Same Thing 13

Chapter 3 The Divisibility Rules:
Hope For the Incurably Lazy 21

Chapter 4 Fractions:
A Safe Bet 37

Chapter 5 Decimals:
Hanging Out at the DeciMall 67

Chapter 6 Percents:
Would you Buy a Used Car From
This Woman? 87

Chapter 7 Averages:
At Home With the Average Family 109

Chapter 8 Ratios:
Things to Do When There's No TV 119

Chapter 9 Units:
The Metric System Won't Go Away 145

Chapter 10 Bases Other than 10:
The Language of Computers 165

Chapter 11 Practice Questions and Answers 177

Glossary 209

About the Author 213

Introduction: Getting to Know Les Ismore

Beauregard lay dozing at the top of the stoop of a brownstone building. He shifted every few minutes in an attempt to keep as much of his body as possible in the sliver of sunlight that passed between the two tall buildings across the street. This was no easy task for a four-foot-tall cat, and Beauregard moved uneasily across the stoop, trying not to get his fur dirty. He thought wistfully of the days of his youth on the porches of South Carolina, where there was plenty of sunlight and a certain lady cat . . . .

"Hey, Beauregard. Quit shoving!"

Beauregard was jolted fully awake by his friend Barnaby, who accompanied his words with a poke from a bony elbow. The cat moved reluctantly out of the sunlight, which shone fully on Barnaby's odd-looking figure. The bright sun on his long, bushy hair and thin frame made him look like a dandelion would look if dandelions wore round glasses and white lab coats. He sat patiently awaiting the arrival of Professor Les Ismore, who was now almost an hour late.

"Please accept my humble apologies," said Beauregard in his least apologetic tone as he winked at Babette, who was also seated on the stoop. Then he settled into the shade behind Barnaby's hair and tried to return his thoughts to South Carolina.

Babette saw Beauregard's wink and she winked back, but her gesture was lost behind her dark glasses. Babette lived in Paris, but came to New York often to visit her friends. They had always admired Babette's sophistication and lovely dark hair, but none of them could remember ever having seen her eyes. Babette also patiently awaited the professor. In fact, according to her relaxed European sense of time he was not late at all, and if he had arrived on time he would have been early.

"It'll be good to see the professor again. But one thing has been bothering me: He's not supposed to be baby-sitting for us, is he?" These last words were spoken by Bridget, who had just returned from the corner store with a fresh supply of bubble gum.

She looked over at Barnaby and blew a perfectly round bubble that was soon large enough to hide not only her face, but also her curly black hair and her New York Yankees baseball cap. She wore the cap backwards, possibly because that kept the bill out of the way of the bubbles she was always blowing.

"Actually," Barnaby replied, "we're supposed to be keeping an eye on him. Remember, Professor Ismore is a mathematical genius, but he's not so good at day-to-day matters."

Hearing this, Beauregard frowned. He already considered it his responsibility to watch over his three companions and the idea of keeping track of a fourth was about as appealing to him as a date with a fox terrier.

"He hasn't been my babysitter for years now," Barnaby continued. "He's been out of town a lot and he always seems to be getting himself in trouble. My mom says that letting him tutor us in math will help keep him out of mischief."

Les Ismore was a friend of Barnaby's parents. They were brilliant, but a bit befuddled, and Barnaby showed every sign of following in their footsteps.

"This sounds like a good idea," said Babette, "the score on my last math test is not something I am proud of."

"Same here," agreed Bridget, "I think the professor is going to have his hands full."

As Bridget spoke, Professor Les Ismore came around the corner and approached the stoop, and as Bridget had predicted, he did in fact have his hands full. He juggled a jumbled mixture that included several bags of groceries and at least a dozen very large books. The bags and books made it impossible for the gang on the stoop to see his face, but they recognized him by his mismatched socks and distinctive waddle. Also, they had never seen him without at least two or three books in his possession.

As the professor came closer, it became clear that the pile of books and groceries not only made it impossible for the gang to see his face, but also made it impossible for the professor to see them. Apparently he also couldn't see the stoop, or anything else that lay beyond the carton of eggs that projected from the grocery bag and pressed up against the tip of his nose.

"Watch out!" yelled Barnaby, but it was too late, and Dr. Ismore tripped on the bottom step of the stoop, pitching books and groceries forward onto our heroes.

It could have been disastrous, but Barnaby and Beauregard were on the top step, and though they were pelted by the fallout from a bag of oatmeal cookies that exploded at their feet, they were safely out of range of the flying books. The lightening-fast reflexes of Babette and Bridget saved them from harm, although the accident cost Bridget one of the better bubbles she had blown in a while and she spent a good part of the next hour picking its remains out of her hair. The professor had pitched forward directly onto his face and escaped harm only through the happy placement of the carton of eggs, which broke his fall in a safe but messy manner.

"Oops, there's egg on my face!" said the professor cheerfully as he sat up, removed a handkerchief from his pocket, and wiped his head free of yolk, white, and shell. After a few swipes of the handkerchief, Dr. Ismore's bright smile, twinkling eyes, and shiny, bald head were revealed.

"Hello! Hello!" he shouted as he recognized his friends. "Here, help me with this stuff and let's go inside."

He raised his pudgy body into a standing position, picked some eggshells off his battered tweed jacket, and stepped carefully up the stairs to open the door while the gang picked up the books and groceries.

Dr. Ismore led them up to the second floor. Once inside, the professor put away the groceries while the gang settled into the book-lined study and ate the surviving oatmeal cookies.

"Wonderful!" said the professor as he entered the room. "It's such a pleasure to see you all again. Bridget, I think this is going to be a good year for your Yankees! Do you remember when we used to go to the games together?"

Bridget smiled, thinking of the time that Dr. Ismore had taken a series of wrong turns during a trip to the snack bar and led them into the Yankee dugout during the fifth inning of a World Series game. The players had been remarkably polite, even autographing a ball for Bridget before the security guards arrived.

"And Babette, you look lovelier than ever. I hope this trip to New York has been a pleasant one and that it will be a lengthy one."

"Thank you, Dr. Ismore," said Babette. "So far, it has been a wonderful trip."

"Excellent. And it will soon get better. Nothing like a little math to turn a good time into a great time."

"Um, if you say so," said Babette, not entirely convinced.

"Barnaby," the professor continued, as he greeted everyone in turn. "Your mother told me about the latest little accident in the lab. A minor matter I'm sure. Don't let it bother you. Some sacrifices must be made if science is to advance and I'm sure that the people in charge of the university will recognize that eventually and let you come back."

"Thanks, Doc," said Barnaby, whose scientific vision sometimes ran too far ahead of his laboratory skills. Barnaby's latest mishap was not actually as bad as some of his previous ones, so he was pretty sure that his present exile from the university was only temporary.

"And Beauregard," said Dr. Ismore, coming around to the cat, who had settled into an armchair by the window. "I was in South Carolina recently and I ran into one of your old acquaintances, a fellow named Old Tim."

"Old Tim!" said Beauregard, suddenly concerned.

"He told me to let you know that the matter that you had wished to remain a secret remains a secret and will continue to remain a secret. He said he hoped that would ease your concern."

"Yes, thank you," said Beauregard. "Although Old Tim is not near the top of the list of cats with whom I enjoy sharing a secret." Beauregard lived a complicated life, and while he served as an excellent caretaker and role model for his companions, he didn't necessarily want them to know everything about his past.

Dr. Ismore looked around the room. Barnaby and Babette were seated at a large oak table in the center of the room. Bridget sat at a desk in the corner in front of the professor's computer and Beauregard continued to occupy the armchair.

"So, gang," said Les Ismore, "I think it's time I share some of my insights about mathematics with you. Before we start, why don't you tell me what you know about numbers."

"I know that one is the loneliest number," said Bridget.

"And thirteen is supposed to be the unluckiest," said Beauregard.

"Eight is enough?" tried Babette.

"On the mean streets of New York," said Barnaby, "you've got to look out for number one."

"And on the dirty sidewalks of Paris, you've got to look out for number two," added Babette.

"At the old ball game it's three strikes and you're out," offered Bridget.

"Hmm, that's not exactly what I had in mind," said the professor. "Where do you think numbers came from?"

"Perhaps the stork brought them?" asked Babette.

"I know," said Bridget. "When two letters love each other very much . . ."

"Let me try again," tried the professor. He was used to having people be patient with him, so he was good at being patient with other people. "How do you think numbers were invented?"

"What do you mean 'invented'?" asked Barnaby. "Haven't numbers always existed?"

"Nope. All right, is everybody comfortable? Have you all settled in?" asked Dr. Ismore.

Everybody answered that they were.

"Okay," said the professor, "let's take a trip. I want to introduce you to a friend of mine. He's the world's first mathematician and he was born about ten thousand years ago." He bent down and picked a book up off the floor.

"This will do," he said, dropping the book on the table.

"Where are we going?" asked Bridget.

"To another time and another place," answered Dr. Ismore.

"How do we get there?" asked Barnaby.

"Easy," said the professor. "Everybody knows that the best way to travel to a different world is to open a book." With that, he opened the book.

Chapter 1
History: Where Do Numbers Come From?

The gang suddenly found themselves standing at the edge of a meadow in the shade of a low cliff. A clear stream trickled by their feet and meandered through the meadow and into a group of trees, and a flock of sheep grazed quietly. Dr. Ismore led everybody toward the entrance to a cave at the base of the cliff.

"Wipe your feet before you come in!" called a voice from inside the cave. "I just cleaned, so if you've got sheep schmutz on your shoes, leave them outside!"

After carefully wiping their feet, the gang entered a tidy little cave that had a small fire burning in one corner. In the cave they saw a little caveman who wore a clean, and neatly-tailored bearskin. His face was well scrubbed and his hair and beard were neatly trimmed.

The professor greeted him. "Og, it's a pleasure to see you again! The sheep look like they're doing well and you look terrific. I see you've been using the comb I gave you."

"Dr. Ismore! Welcome back to prehistory," said Og with a smile and a handshake. "I see you've brought some friends with you."

"Og is the world's first mathematician," said the professor as Og greeted Bridget, Babette, Beauregard, and Barnaby.

"Math is really just a hobby," said Og. "Sheepherding is my real job."

"You speak very good English for a prehistoric sheepherder," said Beauregard.

"Thank you. And you speak very good English for a four-foot-tall cat," replied Og with a grin.

"What can you tell my friends about numbers, Og?" asked Dr. Ismore.

It all started with counting sheep," Og explained. "You see, I can't sleep at night until I've counted all of my sheep. Once I know that they're all safe and sound, I sleep like a baby, but if any of them are missing, I'm awake all night sometimes, looking for them."

"So counting sheep helps you fall asleep?" said Babette.

"Yep, you heard it here first," said Og. "I counted my sheep every night and kept track of them by making notches on the walls of the cave. For a long time, I had this many sheep." Og pointed at a row of notches on the wall.

"But some nights when I counted the sheep, I counted this many notches." Og pointed at another row of notches.

"How many is that?" asked Bridget.

"I can't tell without counting up the rows," said Barnaby.

"The two rows look about the same to me," said Beauregard.

"Well, with all that counting, I bet you weren't getting much sleep," said Babette.

"That was the problem with my system," said Og. "Once I drew more than a couple of notches, it was really hard to tell two different numbers apart and it took me a long time every night to figure out if all the sheep were here. So I came up with an idea that made it a little bit easier."

The professor interrupted. "Og is too modest, his idea became the foundation for every bit of math that has ever happened since."

Og looked embarrassed. "Dr. Ismore is very kind, but I'm really just a humble prehistoric shepherd."

"What was your idea?" asked Bridget.

"Well, instead of just making notches on the wall and counting them one by one, I decided to put the notches in small groups, and then count the groups. Since I counted on my fingers sometimes, I decided to make my notches in groups of five, like hands. So now the number of sheep looked like this:

Now I can look at the wall and see quickly that I have three groups of five, plus two more. I like to say it as 'three hands and two fingers.' Now let's look again at that second row of notches, but this time in groups.

Finger

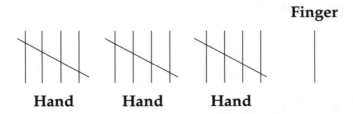

Hand　　　**Hand**　　　**Hand**

I can see right away that there are three hands and only one finger. That means I'm missing a sheep and I have to go out and find it before I can go to sleep."

"By counting numbers in groups, Og invented mathematics," said the professor, slapping Og on the back.

"Around 5000 B.C. the Egyptians took my idea and expanded on it. They came up with different symbols for different sized groups."

"Wait a minute," said Barnaby. "How do you know what happened so far in the future?"

"Well, once word got out that I was the world's first mathematician, I became very popular with mathematicians that were also time travelers. So I know quite a bit about the future."

"Then why are you still counting on your hands and fingers?" asked Barnaby.

"Heck, I've only got seventeen (I mean three hands and two fingers) sheep. Math is fun, but I really like being a prehistoric shepherd and when you spend your day keeping track of seventeen sheep, hands and fingers work as well as any number system I've seen from the future."

"Mathematicians are weird," said Bridget.

"Yeah, so?" said Og and Dr. Ismore at the same time. "Anyway," Og continued, "the Egyptians had different symbols for 1, 10, 100, and 1,000, and this is what they looked like," he said, scratching symbols in the dirt at his feet.

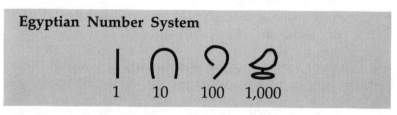

Egyptian Number System

| 1 | 10 | 100 | 1,000 |

Babette took a stick and started scratching symbols in the dirt at her feet. "So tell me if I have this straight. If you counted 17 sheep using the Egyptian system, you'd count 1 ten and 7 ones and it would look like this:

$$\cap \quad \text{||||} = 17$$

"That's right, and here's another one," said Og. "The last time I checked, I had 124 neighbors. How would I write that in Egyptian symbols?"

Bridget answered. "That's 1 hundred, 2 tens, and 4 ones." She took the stick from Babette and drew in the dirt.

$$9 \cap \cap \text{||||} = 124$$

"That's right," said Og.

Now you try a few: (The answers are on page 12.)

✎ ✎ ✎ ✎ ✎

1) Euclid was a Greek mathematician who lived around 2,300 years ago. Not only did he invent geometry, but according to Og, during one visit to prehistory he recited the entire Greek alphabet in the course of a single belch! Euclid's geometry text was called "The Elements" and it contained 13 books. How would the Egyptians have written the number 13?

2) Rene Descartes was perhaps the smartest French person who ever lived. He was born in 1596 and came up with all kinds of ideas in math and philosophy that we still use today. During one visit to Og, he said "I think, therefore I am," which he's famous for saying. He also said, "This wine stinks and this lamb chop is overcooked," which he's not famous for saying. How would the Egyptians have written Descartes' date of birth?

✎ ✎ ✎ ✎ ✎

Og went on. "The Egyptian system was pretty good, but it had its drawbacks. You can see what happened when you had to write numbers with 8s and 9s. Too much work; too many symbols to draw. The Roman system improved on the Egyptian system by adding symbols for 5, 50, and 500. The Roman number system looks like this:

Roman Number System	
I	1
V	5
X	10
L	50
C	100
D	500
M	1,000

"I see," said Beauregard. "Now our 17 sheep can be written as 1 ten, 1 five, and 2 ones, which is a little bit tidier than the Egyptian system." He drew the Roman numbers on the ground with his paw.

XVII = 17

"There's one more thing," Og continued. "The Roman system did something else to keep from having to draw too many symbols. The Romans never drew more than three of any symbol in a row. Instead of writing 4 as IIII, they thought of it as 1 less than 5, and wrote it as IV. Instead of VIIII for 9, they used IX, or 1 less than than 10. So in the Roman system, whenever you see a smaller symbol in front of a bigger symbol, it means you should subtract. Look at these:

4	IV,	1 less than 5
9	IX,	1 less than 10
40	XL,	10 less than 50
90	XC,	10 less than 100
400	CD,	100 less than 500
900	CM,	100 less than 1,000

This way of writing 4s and 9s means that you have fewer symbols, but it makes Roman numbers a little confusing."

"A lot of people like to use Roman numbers for important dates because they look classy," Dr. Ismore pointed out. "You always see Roman numbers on the cornerstones of buildings and at the end of movie credits. Personally, I think 1954 was one of the most important dates in history because that was the year that Elvis Presley recorded his first album."

"Ah, Elvis, one of my favorite mathematicians. I always enjoy his visits." said Og. "You know, musical genius and mathematical genius are almost the same thing."

"So does that mean Elvis was weird, too?" said Bridget.

"The weirdest," said Og. "Now let's write 1954 as a Roman number. It's not so bad if you think of it as 1,000 + 900 + 50 + 4, like this:

1954 = 1,000 + 900 + 50 + 4

M   CM   L   IV = MCMLIV

"Here's another," said Og. "Elvis's favorite food was fried peanut butter and banana sandwiches and there are 949 calories in your average fried peanut butter and banana sandwich. Let's write 949 as a Roman number."

949 = 900 + 40 + 9

CM   XL   IX = CMXLIX

Whew! That was a tough one because all three digits were 4 or 9. Here are a few more for practice.

✎ ✎ ✎ ✎ ✎

3) Sir Isaac Newton invented calculus, an incredibly cool branch of math that you'll probably see in high school. Newton was born in 1642, which makes it an important enough year to try writing it as a Roman number.

4) During his career, Elvis had 36 top-ten hits. Write 36 as a Roman number.

5) Muhammad ibn Musa al-Khwarizmi was one of the greatest mathematicians ever. About the year 830, he wrote a book called "Hisab al-jabr wa'l muqabalah" that described the basics of algebra. In fact, the word algebra comes from al-jabr in the title of the book. How would the Romans write the year 830?

Og went on to say that the number system that we use today was started around the year 600 by the Hindus, in India. The Hindus took a different approach from the Egyptians and Romans. The great thing about the Hindu system was that they had a different symbol for each of the digits 1 through 9. Their symbols were not that different from the ones we still use today. They showed tens, hundreds, and thousands by putting the same nine digits in different parts of the number. So a number like 6,743 really means this:

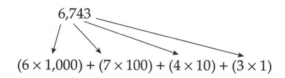

$$(6 \times 1{,}000) + (7 \times 100) + (4 \times 10) + (3 \times 1)$$

The Hindu invention of the ones place, tens place, hundreds place, etc., made addition, subtraction, multiplication, and division possible; just try and multiply two Roman numbers and see how far you get!

While it was a big improvement, the Hindu system did have one of the same drawbacks as the old systems: it had no symbol for zero. After thousands of years of counting, it hadn't occurred to anyone to create a symbol for nothing, and that sometimes made things confusing. For instance, with no symbol for zero, when Og wrote 17 as the number of his sheep, there was no way to tell whether he meant 1 ten and 7 ones, or 1 hundred and 7 tens, or even 1 hundred and 7 ones. So without zero 17, 170, and 107 all look the same.

The confusion over nothing continued for about 300 years while people made gradual changes to try to clear things up. At first, they tried to put spaces where the zeroes would be, so 107 looked like this: 1 7. That wasn't much help, though, so they started putting a dot where a zero would be and that was much better, so now 107 looked like this: 1• 7. Eventually the dot got bigger and became the symbol that we now use for zero (0) around the year A.D. 900.

"So," Og said, "perhaps the single greatest invention in the history of numbers was the invention of nothing."

"Zero, my hero!" said Bridget.

"That's much ado about nothing," said the professor.

"Sweet diddley!" said Beauregard. Diddley is a feline expression that means nothing. That is, it means something, but that something is nothing.

"The Arabs learned Hindu numbers when they visited India, but changed them so that they looked like the ones we use today," Og explained. "They brought them to Europe around the year 1100, and by then, Europeans were pretty tired of Roman numbers, so they took to the new system very quickly. We call the modern number system 'Arabic numbers' because the Arabs taught these numbers to the western world."

"There are also some other number systems that you might want to check out," the shepherd continued. "The Chinese and the Mayans had really cool sets of numbers, but we'll have to leave those for another time. I have a date in a little while."

"With another mathematician from the future?" asked Barnaby.

"Nah, with a shepherd woman from over the hill. How's my hair? Is there any schmutz in my beard?" Og combed his hair.

"You look great," said Babette. "Very well groomed."

"I've met a lot of smart people from the future," said Og. "But I'd like to meet the person who invented this." Og held up the comb. "What do you call this thing, again?"

"A comb," said the professor.

"The comb," repeated Og. "A genius. Whoever invented this is a genius. What it has done for my social life, I can't even begin to tell you. Well, I don't mean to rush away, but I think I see her coming across the meadow. It was nice to meet you all."

"It's been a pleasure," said the professor as he opened the book that would return the gang to the present day. "The next time you see Elvis, tell him I said 'Hello!'"

ANSWERS:

1) 13 = 1 ten and 3 ones.

$$\cap \ |||\ = 13$$

2) 1596 = 1 thousand, 5 hundreds, 9 tens, and 6 ones.

$$= 1,596$$

3) 1642 = 1,000 + 500 + 100 + 40 + 2

              M   D   C   XL   II = MDCXLII

4) 36 = 30 + 5 + 1

         XXX   V   I = XXXVI

5) 830 = 500 + 300 + 30

         D   CCC   XXX = DCCCXXX

Chapter 2
Words and Numbers: They're the Same Thing

"Words and numbers aren't actually all that different, you know," Dr. Ismore pointed out. The group had settled back into the professor's study. Beauregard lay curled up on an Oriental rug and the others were seated around the oak table drinking tea and eating cookies. "It can be very satisfying when you see a bunch of words and you recognize that the words are telling you to do some simple mathematical thing. Words and math, math and words, its really all the same thing."

"I know what you mean," said Bridget. "I can say that I've eaten 14 cookies and Barnaby has had 5 more cookies than I've had. If I want to know how many cookies Barnaby has eaten, the words are telling me to do addition.

Bridget's cookies	14	14
+ 5	+5	+ 5
Barnaby's cookies		19

So if I look at the words carefully, I can see the math that's hidden in them and I know that Barnaby has eaten 19 cookies."

"That's it exactly, " said the professor as he opened up another bag of oatmeal cookies. "Once you can see the connection between words and math, everything becomes much easier and more fun."

"Here's another one," said Bridget. "Let's say the Yankees win 151 games this year. If the baseball season is 162 games long and there are no ties, then how many games will the Yankees lose?" Bridget was a loyal Yankee fan, so it was no surprise that she would choose to have the Yankees win nearly every game.

"You must do subtraction," said Babette.

Total Yankee games	162	162
− Yankee wins	− 151	−151
Yankee losses		11

"151 wins and 11 losses. That would be an awfully good season," said Barnaby.

"All they need is one more left-handed pitcher," insisted Bridget, who was left-handed.

"How about this one," offered Beauregard. "Everybody knows that a cat has nine lives. How many lives do 22 cats have?"

"Ah, multiplication. Excellent," said the professor.

Number of cats	22	22
× 9 lives per cat	× 9	× 9
Total number of lives		198

"There are 120 oatmeal cookies in this bag," said the professor. "If we divide them up evenly among the five of us, how many cookies does each of us get?"

Barnaby giggled. "Well, if we're *dividing* up the cookies, I guess we must be doing division."

Bridget rolled her eyes and said, "You are very easily amused."

Cookies ÷ Number of people = Cookies per person

$$120 \div 5 = 5\overline{)120}$$

with quotient 24, then 12, then 0.

"Twenty-four cookies apiece. That's a lot of cookies," said Beauregard.

"It's okay, they're small," said Barnaby, tossing two more into his mouth.

Okay, reader, why don't you try a few? All you've got to do is turn the words into addition, subtraction, multiplication, or division to find the answers. (The answers are on page 18.)

⊗ ⊗ ⊗ ⊗ ⊗

1) Barnaby is collecting dirty socks for a cutting edge scientific experiment. He needs 104 socks. So far, he has collected 87 socks. How many more socks does he need?

2) Babette has a one-way plane ticket from New York to Paris. The ticket costs $235, but for an extra $55 she can upgrade the ticket to first-class. How much will a first-class ticket cost Babette?

3) A recipe for fish-head stew calls for 6 fish heads per serving. Beauregard needs to cook up enough fish-head stew to serve 14 cats. How many fish heads does he need?

4) Let's stay with Beauregard and his fish heads. If each fish head costs $3 (these are high-quality fish heads), can you use the answer to the previous question to find out how much Beauregard has to pay for fish heads?

5) Bridget has $23, but she owes $6 to Babette. If Bridget pays Babette what she owes her, how much will she have left?

6) Each hair on Barnaby's head is 8 inches long. If Barnaby has 4,680 hairs on his head, how long would Barnaby's hair reach if all of his hairs were laid end to end?

7) There are 240 students at Bridget's school. If there are an equal number of students in each class and there are 15 classes, how many students are there in each class?

8) Babette has 16 pairs of black pants and she has 24 more black shirts than she has pairs of black pants. How many black shirts does Babette have?

9) We found out in question 6 that Barnaby's hair would stretch 37,440 inches if each strand were laid end to end. How long is that in feet? (Remember, there are 12 inches in a foot.)

10) Barnaby is 12 years old and Dr. Ismore is 41 years older than Barnaby. How old is Dr. Ismore?

ANSWERS:

1) This is a subtraction problem.

Total socks he needs	104	104
− Socks he has so far	− 87	− 87
Socks he still needs		17

So Barnaby needs 17 more socks for his experiment.

2) This is addition.

Regular ticket	$235	$235
+ $55 upgrade fee	+ $55	+ $55
First - class ticket		$290

Babette's first-class ticket costs $290. A bargain.

3) Multiplication.

Number of servings	14	14
× Fish heads per serving	× 6	× 6
Total fish heads		84

Beauregard needs 84 fish heads to feed fish-head stew to 14 cats.

4) We know from the previous question that Beauregard needs to buy 84 fish heads. Since we know how much each fish head costs, we can solve the problem by using multiplication.

Number of fish heads	84	84
× Price per fish head	× $3	× $3
Total cost		$252

So the fish-head stew costs $252. Worth every penny, they say.

5) Subtraction.

Money she has at first	$23	$23
− Money she pays	− $6	− $6
Money she has left		$17

Bridget has $17 left.

6) Use multiplication.

Hairs on Barnaby's head	4,680	4,680
× Length of each hair	× 8	× 8
Total length of Barnay's hair		37,440

If all of Barnaby's hairs were laid end to end, they would stretch 37,440 inches.

7) This is a division problem.

Students at the school ÷ Classes = Students per class

$$
\begin{array}{r}
16 \\
15{\overline{\smash{\big)}\,240}} \\
\underline{15} \\
90 \\
\underline{90} \\
0
\end{array}
$$

There are 16 students in each class at Bridget's school.

8) This is addition.

Number of black pants	16	16
+ 24	+ 24	+ 24
Number of black shirts		40

Babette has 40 black shirts.

9) The number of feet will be smaller than the number of inches, so we have to divide.

Length in inches ÷ Inches per foot = Length in feet

$37,440 ÷ 12 =$

$$
\begin{array}{r}
3,120 \\
12\overline{)37,440} \\
36 \\
\hline
1440 \\
1200 \\
\hline
240 \\
240 \\
\hline
0
\end{array}
$$

Barnaby's hair stretches 3,120 feet.

10) This one's addition.

Barnaby's age	12	12
+ 41 more years	+ 41	+ 41
Dr. Ismore's age		53

Dr. Ismore is 53 years old.

Chapter 3

The Divisibility Rules: Hope For the Incurably Lazy

"What if I just want to find out if one number is divisible by another number?" Beauregard wondered. "For instance, what if I had 7,893 chocolate-covered mouse heads and I wanted to know if we could all share them evenly? Isn't there some way to figure that out without going to all the trouble of actually dividing 7,893 by 5?"

Babette, who had started to peel the wrapper off a chocolate bar, stopped and put the candy back in her pocket. Even her great love of chocolate was ruined at the thought of biting into a chocolate-covered mouse head.

"Ah, of course!" cried the professor, who didn't seem at all offended by Beauregard's taste in snacks, "you mean the divisibility rules! If you know how to use the divisibility rules, you can sometimes avoid doing long division. A wonderful shortcut! A lazy person's best friend!"

"I would prefer to think of myself as efficient," sniffed Beauregard, who thought he had been insulted.

"Nonsense, and it's nothing to be ashamed of!" replied the professor. "All of the great advances in civilization have been the result of laziness. Brilliant ideas are usually the result of an attempt to do things more easily and quickly. Why, eventually we'll be able to do everything so easily and quickly that we won't do anything at all." The gang looked at each other quizzically. "Well, never mind that," continued the professor, "I know just the man to teach us the divisibility rules, my old friend Mycroft, and he lives right downstairs."

In a few minutes the gang was seated on an overstuffed couch in Mycroft's apartment. Mycroft sat facing them in an overstuffed chair. He wore an elegant looking pair of pajamas and he rested his feet on an overstuffed pillow.

"May I offer you some refreshments?" he asked, after greeting everyone in turn without rising from the chair. In fact, he had managed to let them in, settle them on the couch, and make them feel welcome while barely moving at all. "Instant coffee? Powdered juice? Soup-in-a-bag? Of course, you'll have to make it yourself, but help yourselves to anything you'd like."

"Thank you," said Barnaby as he rose and went to the kitchen. He started mixing powders and fluids in containers, all of which he found within easy reach.

"I believe that Mycroft is uniquely qualified to help us with the divisibility rules," said the professor with a smile, "because he is the laziest person I know."

"I prefer to think of myself as efficient," Mycroft replied. The kids all looked at Beauregard, who smiled smugly.

"Well, let's get started," Mycroft continued, " I'll be happy to help you with divisibility as long as it doesn't involve moving from my chair or doing long division. First, what is divisibility?"

"A number is **divisible** by a second number when it can be divided by the second number without leaving **a remainder**," said Babette. "So 9 is divisible by 3 because 9 can be divided by 3 with no remainder, but 5 is *not* divisible by 2 because there is a remainder when 5 is divided by 2; the remainder is 1." A remainder is what's left over when two numbers can't be divided evenly. Two goes into 5 twice, with 1 left over, so the remainder is 1.

"That's pretty good," said Mycroft. "Numbers are divisible when they can be divided without too much trouble. No muss, no fuss, no remainders, no fractions, no decimals, none of those things that make math start to feel like work."

Hey, reader! Try a few easy ones (the answers are on page 33):

1) Is 11 divisible by 3?

2) Can 12 hot dogs be shared equally by 4 people without cutting any of the hot dogs?

3) Barnaby, Beauregard, Babette, Bridget, Dr. Ismore, and Mycroft are sitting in Mycroft's study. Can they be split into two equal groups?

"Okay, now we know what divisibility is, and that's all very nice," said Beauregard. "But what if the numbers are really big. What about my 7,893 chocolate-covered mouse heads?"

"For bigger numbers, we need to use the divisibility rules," replied Mycroft. "The divisibility rules are a set of tricks that help you tell if a big number is divisible by a number between 1 and 10 without actually doing the long division. Let's start with 1, because that's the easiest number. How can you tell if a number is divisible by 1?"

"Well, duh," Bridget answered, sarcastically, "all numbers are divisible by 1!"

"Excellent!" Mycroft's reply was enthusiastic, although he didn't raise his voice or seem to move in any way. "That's our first divisibility rule. All whole numbers are divisible by 1. Just because it's easy doesn't mean it isn't important!"

4) Beauregard offered his 7,893 chocolate-covered mouse heads to each of his friends and they all turned him down. He then offered to share the mouse heads with all of his acquaintances, who also refused. He even asked strangers on the street if they wanted any, but nobody did. So Beauregard set out to eat them by himself. The question is: can 7,893 be divided evenly by 1?

"How about 2?" continued Mycroft. "Does anybody know the divisibilty rule for 2?"

"That's pretty easy," said Barnaby as he returned from the kitchen carrying a tray that held various murky fluids in bowls and glasses. "All even numbers are divisible by 2. So if a number ends in 2, 4, 6, 8 or 0, it's divisible by 2.

5) Just as Beauregard sat down to eat the 7,893 mouse heads, he was joined by his feline friend Luigi. Luigi was hungry and the two cats decided to split the mouse heads. Can they split them evenly?

6) Here's the $64,000 question: Can 64,000 be divided evenly by 2?

"Please help yourselves," Barnaby offered as he put the tray on a table.

"Um, what are our choices?" asked Bridget suspiciously.

"The greenish stuff is decaf split pea soup and the brownish stuff is french onion fruit punch," said Barnaby. "I thought I'd make it interesting."

"No, thank you," said Bridget.

"Perhaps later," said Babette.

"I had a big breakfast," said Beauregard.

"I'll pass," said Mycroft.

"Don't mind if I do," said the professor as he grabbed a glass of each of Barnaby's creations, casually poured the two together, and drank the greyish purple mixture down in two gulps. "Not bad, my compliments to the chef," he said, wiping his mouth while the others squirmed and Barnaby smiled proudly.

"How do we tell if a number is divisible by 3?" asked Bridget. "That's one I don't know."

"Ah!" said Mycroft. "That's my favorite divisiblity trick. To tell if a number is divisible by 3, you add up the digits, and if their sum is divisible by 3, then the original number is divisible by 3."

"Eh?" said Babette. ("Eh?" is French for "Huh?")

"Let me show you an example," said Mycroft.

"Let's see if 486 is divisible by 3:
Add up the digits:

$$4 + 8 + 6 = 18$$

The sum of the digits, 18, is divisible by 3, so the original number, 486, must also be divisible by 3.

Try another one. Is 2,604 divisible by 3?

Add up the digits:

$$2 + 6 + 0 + 4 = 12$$

The sum of the digits, 12, is divisible by 3, so the original number, 2,604, is divisible by 3.

One more. Is 373 divisible by 3?

Add up the digits:

$$3 + 7 + 3 = 13$$

The sum of the digits, 13, is not divisible by 3, so the original number, 373, is not divisible by 3."

Okay, reader, now you try a few.

✎ ✎ ✎ ✎ ✎

7) A book is 114 pages long. Can the book be divided into three chapters, each of which has the same number of pages?

8) Barnaby has a drawer containing 220 socks and he wants to know whether he can divide them into groups of three with none left over. He can think of no practical reason for doing this, but he wants to do it in the interest of pure scientific speculation. Can he do it?

9) Let's get back to the chocolate-covered mouse heads. If Beauregard and Luigi are joined by their friend Rufus, can the three cats divide the 7,893 mouse heads evenly?

10) Mycroft has 621 packets of instant soup in his kitchen. If he has split pea, chicken noodle, and french onion, could there be equal numbers of packets of the three types of soup in his kitchen?

✎ ✎ ✎ ✎ ✎

"Does anybody know the divisibility rule for 4?" asked Mycroft. "I could explain it, but I'd prefer to rest for a little while."

Bridget jumped up, excited. "I know that one! That one is my favorite. All you have to do to tell if a number is divisible by 4 is look at the last two digits. If the last two digits are a number that is divisible by 4, then the entire number is divisible by 4. So I know that 3,868,716 is divisible by 4 because the number fomed by the last two digits, 16, is divisible by 4."

Babette looked troubled. "But the rest of the number is not important?"

"Nope," replied Bridget with absolute confidence, "all you have to do is look at the last two digits. The rest of the digits don't make any difference at all. So 312; 1,812; and 6,987,675,912 are all divisible by 4 because they all end in 12. Twelve is divisible by 4, and that's the only thing we care about!"

"So 20; 120; 220; 320; and 420 are all divisible by 4 because they all end in 20?" Babette asked, who was starting to get the idea.

"Exactly!" said Bridget.

Here are some questions about divisibility by four.

11) Bridget, Babette, Beauregard, and Barnaby have found a bag containing $648, all in one dollar bills. Can they divide the dollar bills evenly among themselves?

12) Is 234,654,987,678,867,580 divisible by 4?

13) More mouse heads. Can the 7,893 mouse heads be divided evenly among 4 cats?

"I can tell you guys the divisibility trick for 5," said Barnaby in a whisper. He whispered because Mycroft seemed to have fallen asleep in his chair. "A number is divisible by 5 when it ends in a 5 or a 0."

Everybody agreed that the rule for 5 made perfect sense. After all, the multiples of 5 are 5, 10, 15, 20, 25, etc, and the 5, 0, 5, 0 pattern just repeats itself over and over again.

✎ ✎ ✎ ✎ ✎

14) Can 435 people be divided evenly into groups of 5?

15) Is 554 divisible by 5?

✎ ✎ ✎ ✎ ✎

"Is it time to talk about divisibility by 6 yet?" asked Mycroft, starting awake in his chair. "The rule for divisibility by 5 is easy enough that I thought you could handle it without my help. The rule for 6 is a little more interesting."

Mycroft then explained to the group that dividing by 6 is the same as dividing by 2 and then dividing by 3. For instance:

$$48 \div 6 = 8 \text{ or}$$
$$48 \div 2 = 24 \text{ and } 24 \div 3 = 8$$

Notice that you get 8 as the answer if you divide by 6, and you get 8 if you divide by 2 and then by 3. That means that if a number is divisible by 6, then it must be divisible by 2 and then by 3. So the test for divisibility by 6 is to do BOTH the test for 2 and the test for 3. For instance, 522 is divisible by 2 because it is even, so it passes the first test for divisibility, AND you get 9 when you add up the digits, so it passes the test for 3. So 522 is divisible by 6 because it is divisible by BOTH 2 and 3.

Let's see if 3,714 is divisible by 6:

3,714 is even, so it's divisible by 2.

Now add up the digits:

3 + 7 + 1 + 4 = 15; 15 is divisible by 3, so 3,714 is also divisible by 3.

So 3,714 is divisible by 6 because it passes BOTH the test for divisibility by 2 and by 3.

If a number passes only one of the tests and not the other, then it is not divisible by 6. For instance, 62 passes the test for 2, but not the test for 3, so it is not divisible by 6, and 63 passes the test for 3, but not the test for 2, so it isn't divisible by 6 either.

Try a few:

✎ ✎ ✎ ✎ ✎

16) Let's get the mouse heads out of the way first. Can 7,893 mouse heads be divided evenly among 6 cats?

17) Barnaby is one of 126 people who are driving to a scientific convention. Can they fit in cars so that there are exactly six people in each car? (These are pretty big cars.)

18) If the six people in Mycroft's study each do an equal number of math problems, can they do exactly 858 problems? (Beauregard isn't really a person, but he acts enough like one that we'll just call him one for the sake of this question.)

✎ ✎ ✎ ✎ ✎

"Now I owe all of you an apology," Mycroft said sadly. "You see, whoever said that 7 was a lucky number wasn't talking about divisibility."

"So the divisibility trick for 7 is difficult?" asked Babette.

Mycroft sighed. "Worse than that. There is no divisibility trick for 7. If you want to know if a number is divisible by 7, the only way to tell is to divide by 7. Long division. Hard labor. Sweat. Calluses. I'm exhausted just thinking about it. Let's move on to divisibility by 8 while I still have a little bit of energy."

"If I remember correctly," said Dr. Ismore, "the test for divisibility by 8 isn't so great either."

Mycroft sighed again. "That's true. The test for divisibility by 8 is kind of like the test for divisiblity by 4, only it's not as good. In the test for 4, you look at the last TWO digits, but in the test for 8 you have to look at the last

THREE digits. So if the last three digits of some big number are divisible by 8, then the entire number is divisible by 8. The big problem with the test for 8 is that you still have to do long division to see if your three-digit number is divisible by 8, but at least this trick might save you some time with a really big number. For instance, if you want to know if 7,987,456,216 is divisible by 8, you can find out by dividing 216 by 8. Look:

$$
\begin{array}{r}
27 \\
8{\overline{\smash{\big)}\,216}} \\
\underline{16} \\
56 \\
\underline{56} \\
0
\end{array}
$$

Eight goes into 216 evenly, so 7,987,456,216 must be divisible by 8."

"That still looks quite a bit like long division," observed Beauregard.

"That's true," admitted Mycroft. "But at least it's *less* long division."

Let's try another. Is 45,362 divisible by 8?

Just use the last three digits in the long division.

$$
\begin{array}{r}
45 \\
8{\overline{\smash{\big)}\,362}} \\
\underline{32} \\
42 \\
\underline{40} \\
2
\end{array}
$$

Eight doesn't go into 362 evenly, so 45,362 is not divisible by 8.

After doing a couple of these, Mycroft and the gang agreed that they weren't much fun, so we'll let you off the hook and not make you do any of these on your own. Just remember that you can save a little time when you check divisibility by 8 if you use only the last three digits of the number that you're checking.

Six, seven, and eight had left Mycroft almost completely worn out. The kids, cat, and professor were also pretty tired. Fortunately the divisibility tests for 9 and 10 were nice and easy. Barnaby explained the divisibility test for 9: "It's almost exactly like the test for 3, except that the digits have to add up to a number divisible by 9. So if the digits of a number add up to a multiple of 9, then the original number is divisible by 9. I'll show you one. Is 4,833 divisible by 9?

Add up the digits:

$$4 + 8 + 3 + 3 = 18$$

18 is divisible by 9, so 4,833 is divisible by 9.
Let's do another: Is 933 divisible by 9?
Add up the digits:

$$9 + 3 + 3 = 15$$

15 is not divisible by 9, so 933 is not divisible by 9. Try a couple on your own:

✎ ✎ ✎ ✎ ✎

19) Once more with the mouse heads. If Beauregard decides to spread the eating of the 7,893 chocolate-covered mouse heads over his nine lives, can he eat the same number of mouse heads in each of his lives?

20) If there are nine players on a baseball team, can 126 kids be divided into baseball teams so that every team has exactly nine players and no kids are left out?

✎ ✎ ✎ ✎ ✎

Finally, Bridget pointed out that any number that is divisible by 10 has to end in a zero. So 670 is divisible by 10, but 672 isn't. Everyone agreed that this was the most user-friendly of the divisibility rules, and they also agreed that it was time to leave Mycroft to his peace and his pillows.

"I've enjoyed our time together, my young friends. Please accept my apologies for not getting up and showing you to the door," Mycroft called cheerily as the group rose to go back upstairs.

Here's a handy roundup of the divisibility rules:

1   All whole numbers are divisible by 1.

2   All even numbers are divisible by 2.

3   Add up the digits. If the sum of the digits is divisible by 3, then the number is divisible by 3.

4   Look at the last two digits. If the last two digits are a number that's divisible by 4, then the entire number is divisible by 4.

5   If a number ends in 0 or 5, then it's divisible by 5.

6   If a number is divisible by BOTH 2 and 3, then it's divisible by 6.

7   There's no trick. You've got to do long division to see if a number is divisible by 7.

8   Look at the last three digits. If the last three digits are a number that's divisible by 8, then the entire number is divisible by 8.

9   Add up the digits. If the sum of the digits is divisible by 9, then the number is divisible by 9.

10 If a number ends in 0, then it's divisible by 10.

ANSWERS:

1) No, 11 is not divisible by 3 because when 11 is divided by 3 there is a remainder of 2.

2) Yes, 12 hot dogs can be shared equally by 4 people. That's just another way of saying that 12 is divisible by 4.

3) Yes, the 6 people in the room can be split into 2 equal groups. So 6 is divisible by 2.

4) Of course 7,893 is divisible by 1; all numbers are divisible by 1.

5) No, 7,893 is an odd number, so it is not divisible by 2.

6) Yes, 64,000 is an even number, so it is divisible by 2.

7) That's just another way of asking if 114 is divisible by 3.
   Add up the digits:

   $$1 + 1 + 4 = 6$$

   The sum of the digits, 6, is divisible by 3, so the number of pages, 114, is also divisible by 3.

8) The question is: Is 220 divisible by 3?
   Add up the digits:

   $$2 + 2 + 0 = 4$$

   The sum of the digits, 4, is not divisible by 3, so the 220 socks cannot be divided evenly into groups of three.

9) Is 7,893 divisible by 3?
   Add up the digits:

   $$7 + 8 + 9 + 3 = 27$$

   The sum of the digits, 27, is divisible by 3, so the three cats can divide up the 7,893 mouse heads evenly.

10) Is 621 divisible by 3?

Add up the digits:

$$6 + 2 + 1 = 9$$

The sum of the digits, 9, is divisible by 3, so there can be equal numbers of packets of the three types of instant soup.

11) Look at the last two digits. The number represented by the last two digits, 48, is divisible by 4, so the entire number, 648, is divisible by 4.

12) It doesn't matter how big the number is, all you have to do is look at the last two digits. The number represented by the last two digits, 80, is divisible by 4, so the entire number, 234,654,987,678,867,580, is divisible by 4.

13) The number represented by the last two digits, 93, is not divisible by 4, so the mouse heads cannot be divided evenly among 4 cats.

14) Yup, 435 ends in a 5, so it's divisible by 5. That's all there is to it.

15) Nope. The number 554 doesn't end in a 0 or a 5, so it's not divisible by 5.

16) 7,893 passes the 3 test because its digits add up to 27, which is divisible by 3, but 7,893 is an odd number so it fails the 2 test. So 7,893 mouse heads can not be divided evenly among 6 cats.

17) Yes, 126 is even, so it passes the 2 test. The digits in 126 add up to 9, so it passes the 3 test. So 126 people can travel exactly six to each car.

18) This question asks whether 858 is divisible by 6; 858 is even, so it passes the 2 test, and its digits add up to 21, so it passes the 3 test; 858 passes both the 2 and the 3 test, so 858 is divisible by 6. That means that the six people in the study can do exactly 858 math problems if they each do the same number.

19) The question is: Is 7,893 divisible by 9?
Add up the digits:

$$7 + 8 + 9 + 3 = 27$$

27 is divisible by 9, so 7,893 is divisible by 9 and Beauregard can eat the same number of mouse heads in each of his nine lives.

20) Is 126 divisible by 9?
Add up the digits:

$$1 + 2 + 6 = 9$$

The digits add up to 9, so 126 is divisible by 9 and the players can be divided evenly.

Chapter 4
Fractions: A Safe Bet

The professor took a deck of playing cards from the drawer of an old, wooden, rolltop desk and sat down at the table, saying, "Have a seat folks, and, Barnaby, bring an extra chair to the table." The gang gathered around him while he shuffled and reshuffled the cards. "Observe carefully. Nothing up my sleeve," he said, spreading the cards out in front of him. With expert hands, Dr. Ismore then dealt five cards, face down, to each person at the table. He also dealt five cards to the space at the table in front of the empty chair.

"Are we going to gamble?" asked Babette.

"We're going to talk about **fractions**," replied Dr. Ismore. With that, he tapped the remainder of the deck, and a card appeared between his thumb and index finger. He spun the card across the back of his hand, revealing it to be the jack of hearts, then he winked and tossed the card at the empty chair opposite him. In a moment, the chair was no longer empty.

"One-eyed Jack, at your service," said the man who now occupied the chair. His handlebar mustache and handsome, angular features did in fact make him look an awful lot like the playing card. He wore a string tie, ruffled shirt, and black Stetson hat, and had a black patch over his left eye. "Les Ismore! What brings you and your friends out to my muddy Mississippi River home?"

The kids now looked around and realized that One-eyed Jack had not been transported into the professor's study. Rather, the entire group was now seated around a poker table in what appeared to be a fancy casino. The gentle up-and-down motion of the entire room made it clear that the casino was in a riverboat. The words of their host made it clear that the riverboat was gently bobbing on the waters of the Mississippi River.

"Aren't you going to introduce me to your friends?" Jack drawled.

"Certainly," said the professor. "Jack, these are my friends Barnaby, Babette, Bridget, and Beauregard. They've been looking after me lately, keeping me out of trouble."

"Not an easy task, if I remember correctly," said Jack with a wicked smile.

"Ah, I was younger then," said Dr. Ismore. "Barnaby, Babette, Bridget, and Beauregard, this is One-eyed Jack, one of the great riverboat gamblers, and a dear friend from my misspent youth."

Jack reached across the table, shaking hands with everybody in turn. He seemed struck by something as he shook Beauregard's paw. "You remind me of someone I used to know," he said. "You wouldn't by any chance be familiar with a good old South Carolina cat by the name of Branford?"

"Ah, Uncle Branford! He used to do card tricks for us when we were kittens," Beauregard said wistfully. "We all loved him. Too bad about what happened. Mother always said that he would come to a bad end."

"Yessirree, if it weren't for that nasty little streak of dishonesty, he would have been as fine a tabby cat as ever anted, bet, and bluffed. I never did hear the full story, but they say when a cat gets caught cheating in a border town, not even nine hundred lives are enough."

"It's a lesson for all of us. Cheaters don't win, winners don't cheat," said Beauregard.

"And you should know your fractions," said Jack. "If old Branford had spent a little more time practicing fractions and a little less time hiding aces in his fur, things might have been different. But enough about the past. Let's talk about fractions."

Jack spread a deck of cards face up on the table. The gang looked at the 52 cards spread out according to the four suits: hearts, diamonds, clubs, and spades. Each suit had 13 cards: ace, two, three, all the way up through ten, jack, queen, and king.

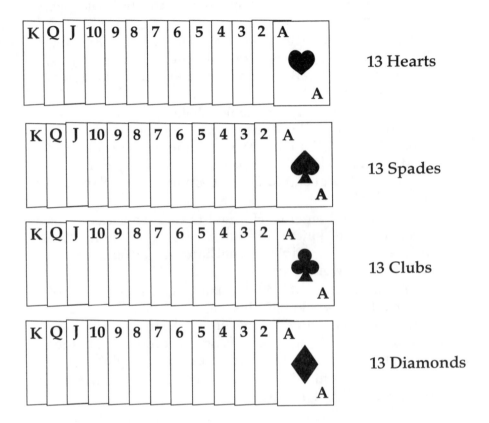

K	Q	J	10	9	8	7	6	5	4	3	2	A

13 Hearts

13 Spades

13 Clubs

13 Diamonds

"If I take all 4 kings, what fraction of the cards have I taken?" Jack began.

Bridget answered, "A fraction is a part over a whole, like this:

$$\text{Fraction} = \frac{\text{part}}{\text{whole}}$$

The part of the deck that you've taken is 4 cards, and the whole deck has 52 cards," Bridget said, separating the kings from the other cards.

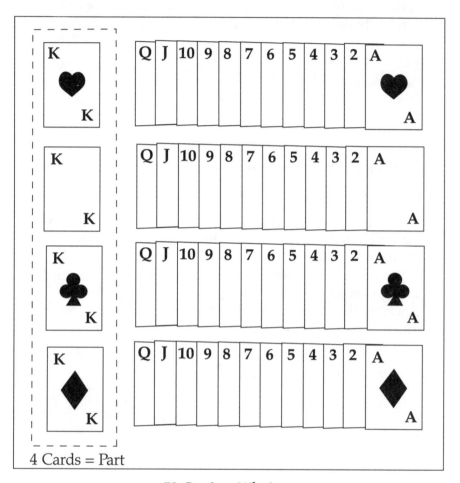

4 Cards = Part

52 Cards = Whole

"So the fraction looks like this:

$$\frac{\text{part}}{\text{whole}} = \frac{4}{52}$$

But you can make the fraction simpler," Bridget explained. "You can divide the top and bottom of a fraction by the same number without changing its value. Both the top and bottom numbers in our fraction, 4 and 52, are even, so they're both divisible by 2:

$$\frac{4 \div 2}{52 \div 2} = \frac{2}{26}$$

But it looks like we can divide by 2 again:

$$\frac{2 \div 2}{26 \div 2} = \frac{1}{13}$$

That's as simple as the fraction can get. So now we know that 4 kings make up $\frac{1}{13}$ of a deck of cards."

"Try another," said Jack. "We know that hearts and diamonds are red and spades and clubs are black. What fraction of the cards are black?"

"That's not bad," said Bridget as she looked at the cards still spread out on the table. "There are 13 spades and 13 clubs, so that's 26 black cards out of a total of 52. So the fraction looks like this:

$$\frac{\text{part}}{\text{whole}} = \frac{26}{52}$$

If you notice that 26 is half of 52, you can reduce this one, too.

$$\frac{26}{52} = \frac{1}{2}$$

So half of the cards are black and the other half are red."

"By the way," said the professor, "the top and bottom numbers in a fraction have names. The top number is called the **numerator** and the bottom number is called the **denominator**. Also, some people use fractions as a way to show division. Think about the number $\frac{1}{2}$. If you have half of something, it's the same as saying you have one thing that's been divided into two. So $\frac{1}{2}$ is the same as

$1 \div 2$. Any fraction is the same as division, so:

$$\frac{\text{anything}}{\text{anything else}} = \text{anything} \div \text{anything else."}$$

"That's not all we know," said Jack. "Fractions can also tell us about **probability**. Not only do we know that 4 kings make up $\frac{1}{13}$ of the deck, we also know that if we were to pick a card at random, the probability of picking a king would be 1 out of 13, or $\frac{1}{13}$. That's a mighty useful tool for a gambler. Let's try another one. If I pick a random card from the deck, what's the probability that I'll pick a heart?"

Barnaby spread the cards on the table and separated out the hearts.

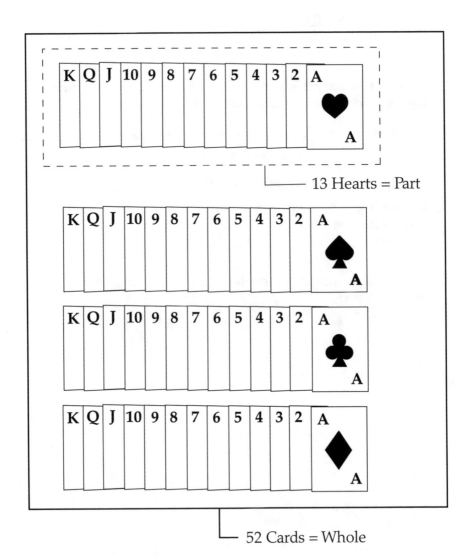

13 Hearts = Part

52 Cards = Whole

"There are 13 hearts in the deck, so that's the part. The whole deck is still 52 cards.

$$\frac{\text{part}}{\text{whole}} = \frac{13}{52}$$

But that's not the simplest form of the fraction, because we can divide the numerator and the denominator by 13.

$$\frac{13 \div 13}{52 \div 13} = \frac{1}{4}$$

So $\frac{1}{4}$ of the cards in the deck are hearts and the probability of picking a heart out of a shuffled deck is $\frac{1}{4}$. That's the same as saying that about 1 out of every 4 cards you pick will be a heart."

"A probability is always a fraction between 0 and 1," said Jack. "The closer to 1 the fraction gets, the greater the probability. If the probability of an event is equal to 1, then the event is a sure thing. The probability of picking a heart, spade, club, or diamond out of a deck is $\frac{52}{52}$, or 1. That means that if you pick a card out of a deck, you are guaranteed to get either a heart, spade, club, or diamond (as long as there are no jokers!).

"If the probability of an event is equal to 0, then the event will never happen. The probability of reaching into a regular deck of cards and getting the Old Maid (she's from a different kind of card game) is $\frac{0}{52}$, or 0, because there is no Old Maid in a regular deck of cards."

"Now," said Jack with a twinkle in his eye, "which is the better bet: When I reach into the shuffled deck, am I more likely to pick a heart or a king?"

"A heart, of course," answered Babette. The probability of picking a heart is $\frac{1}{4}$, while the probability of picking a king is only $\frac{1}{13}$. Now, $\frac{1}{4}$ is bigger than $\frac{1}{13}$ and a bigger fraction means a higher probability.

"Hey," said Barnaby, "What if it's not so easy to tell which fraction is bigger? Like, what if I need to know which is bigger: $\frac{2}{5}$ or $\frac{3}{7}$?"

"Then you use the bowtie," replied Jack.

"You mean the thing around your neck?" said Barnaby.

"No, that's a string tie. This is a bowtie," Jack said, taking out a pencil and a scrap of paper. "First you write the fractions side by side:

$$\frac{2}{5} \qquad \frac{3}{7}$$

Then draw the bowtie. That's the two arrows pointing upward.

$$\frac{2}{5} \diagup\!\!\!\!\diagdown \frac{3}{7}$$

Now multiply the numbers along the arrows and put the products at the end of the arrows above the fractions.

$$\overset{14}{\frac{2}{5}} \diagup\!\!\!\!\diagdown \overset{15}{\frac{3}{7}}$$

The bigger number is always above the bigger fraction. We know $\frac{3}{7}$ is bigger than $\frac{2}{5}$ because 15 is bigger than 14."

Now you try a few problems (the answers are on page 58).

✎ ✎ ✎ ✎ ✎

1) Jack, Dr. Ismore, Babette, Bridget, Beauregard, and Barnaby are sitting around a table. What fraction of those present at the table are female?

2) What is the probability of reaching into a deck of cards and getting a "face card"? Face cards are jacks, queens, and kings.

3) Beauregard threw a party that was attended by 5 cats (including Beauregard), 3 dogs, and a bullfrog (his name was Jeremiah, he made the punch). What fraction of the animals at the party were dogs?

4) Barnaby has collected 8 gallons of horsesweat for an experiment. If he uses 2 gallons of it in a lemonade recipe, what fraction of the horse sweat has he used?

5) Babette has read $\frac{5}{8}$ of the book *Tom Sawyer* and Bridget has read $\frac{7}{9}$ of the same book. Who has read the greater fraction of the book?

6) At the dinner table, Barnaby takes $\frac{1}{3}$ of the mashed potatoes and Babette takes $\frac{3}{10}$. Who has taken the greater fraction of the mashed potatoes?

✎ ✎ ✎ ✎ ✎

"What if we need to add fractions?" asked Babette. "What if I climb $\frac{1}{3}$ of the way up the Eiffel Tower before noon and then another $\frac{2}{5}$ of the way up after noon? How much of the way have I climbed?"

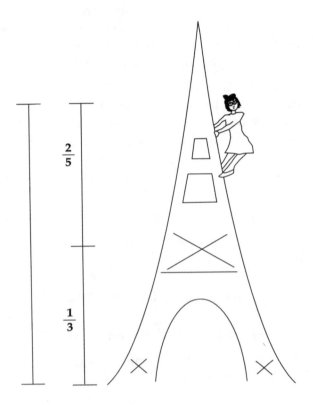

Jack answered, "The important thing to remember when you're adding fractions is that you first have to make the two denominators the same. Remember that the denominator is the number on the bottom. Mathematicians call this 'finding a **common denominator**.' Once you make the bottom numbers the same, adding fractions isn't too tough. It turns out that finding a common denominator isn't hard either, all you have to do is multiply the two denominators that you have.

Let's take it step by step. First write out the question:

$$\frac{1}{3} + \frac{2}{5} =$$

Now multiply the numbers on the bottom to get a common denominator.

$$\frac{1}{3} \xrightarrow{+} \frac{2}{5} = \frac{}{15}$$

To get the top of the new fraction, we use the bowtie.

$$\overset{5}{} \quad \overset{6}{}$$
$$\frac{1}{3} \bowtie \frac{2}{5} = \frac{}{15}$$

Now we add the two numbers we got from the bowtie to get our answer."

$$\frac{1}{3} + \frac{2}{5} = \frac{5 + 6}{15} = \frac{11}{15}$$

"So Babette has climbed $\frac{11}{15}$ of the way up the Eiffel Tower," said Bridget. "Let's try another one. Remember that awful experiment that Barnaby did last year?"

"You mean the one where he burned $\frac{1}{4}$ of his laboratory?" asked Beauregard.

"Right," said Bridget. "Then he destroyed another $\frac{1}{3}$ of the lab with the flood he caused trying to put the fire out. What fraction of the lab did he destroy, anyway?" First write out the question.

$$\frac{1}{4} + \frac{1}{3} =$$

Now multiply across the bottom.

$$\frac{1}{4} \xrightarrow{+} \frac{1}{3} = \frac{}{12}$$

Do the bowtie.

$$\overset{3}{} \quad \overset{4}{}$$
$$\frac{1}{4} \bowtie \frac{1}{3} = \frac{}{12}$$

Add the numbers you got from the bowtie to get the numerator of the answer.

$$\frac{1}{4} + \frac{1}{3} = \frac{3 + 4}{12} = \frac{7}{12}$$

So last year Barnaby destroyed $\frac{7}{12}$ of his laboratory.

"One of the nice things about this method is that it works just as well for subtraction," said Jack. "You follow exactly the same process for subtraction except that you subtract the two numbers you get from the bowtie instead of adding them."

"Here's one," said the professor. "Beauregard and his friends were out catting around the other night, and when they first went out, they had $\frac{1}{2}$ pound of catnip."

Beauregard looked sheepish, which is not easy for a cat.

"When they came back at dawn, they had $\frac{3}{16}$ of a pound. How much catnip did they use up during their carousing?"

First set up the question.

(Catnip they started with) – (Catnip left at the end) = (Catnip they used up)

$$\frac{1}{2} - \frac{3}{16} =$$

Multiply across the bottom.

$$\frac{1}{2} \longrightarrow \frac{3}{16} = \frac{}{32}$$

Do the bowtie.

$$\overset{16}{\underset{2}{\frac{1}{}}} \times \overset{6}{\underset{16}{\frac{3}{}}} = \frac{}{32}$$

Now, instead of adding the numbers above the fractions, we subtract.

$$\frac{1}{2} - \frac{3}{16} = \frac{16 - 6}{32} = \frac{10}{32}$$

We can reduce $\frac{10}{32}$ further because the numerator and denominator are both divisible by 2.

$$\frac{10 \div 2}{32 \div 2} = \frac{5}{16}$$

"So Beauregard and his friends used up $\frac{5}{16}$ of a pound of catnip the other night," said the professor.

"Rest assured, we nipped responsibly," said Beauregard.

Hey, reader! Try using the bowtie to solve these addition and subtraction problems.

✎ ✎ ✎ ✎ ✎

7) Barnaby has an idea, but it's only $\frac{1}{2}$ baked. If he thinks about it a little longer and it becomes another $\frac{3}{7}$ baked, how well baked will his idea be?

8) Babette eats $\frac{2}{5}$ of a pie and Bridget eats another $\frac{2}{9}$ of the pie. What fractional part of the pie have they eaten together? You didn't think you were going to get through a chapter about fractions without hearing about pies, did you?

9) The probability of drawing a spade out of a deck of cards is $\frac{1}{4}$. The probability of rolling a seven with a pair of dice is $\frac{1}{6}$. How much greater is the probability of drawing a spade than the probability of rolling a seven?

10) Bridget picked up the sports page and pointed out that ace pitcher Lefty Lizzouli was responsible for winning $\frac{2}{7}$ of the games won by the Yankees last year. If Righty Randazzo was responsible for winning $\frac{1}{5}$ of the games, what fractional part of the games won by the Yankees last year were won by Lefty Lizzouli or Righty Randazzo?

11) Dr. Ismore had a bottle of mouthwash that was $\frac{3}{4}$ full. He drank some of it by mistake and when he finished he saw that the bottle was $\frac{1}{5}$ full. How much mouthwash did the professor drink?

✐ ✐ ✐ ✐ ✐

"What about multiplication?" asked Barnaby. "What if Dr. Ismore's bottle of mouthwash was $\frac{3}{4}$ full and he drank $\frac{1}{3}$ of what was there. Now how much is left in the bottle?"

"Good question," said Jack. "When you see the word "of" in a math problem, that means you should multiply. So if he drank $\frac{1}{3}$ of $\frac{3}{4}$ of the bottle, it's the same thing as $\frac{1}{3} \times \frac{3}{4}$. Once you've set up a multiplication problem, the rest of the work isn't too hard because all you have to do is multiply straight across the top and straight across the bottom to get your answer."

Take a look.

$$\frac{1}{3} \times \frac{3}{4} = \frac{1 \times 3}{3 \times 4} = \frac{3}{12}$$

The top and bottom of our answer are both divisible by 3, so we can simplify.

$$\frac{3 \div 3}{12 \div 3} = \frac{1}{4}$$

Dr. Ismore has drunk $\frac{1}{4}$ of the bottle of mouthwash.

"So $\frac{2}{3}$ of $\frac{7}{8}$ is the same as $\frac{2}{3} \times \frac{7}{8}$?" asked Babette.

"That's right," answered Jack. "And by multiplying straight across the top and bottom, we get

$$\frac{2}{3} \times \frac{7}{8} = \frac{2 \times 7}{3 \times 8} = \frac{14}{24}.$$

Then we can divide top and bottom by 2 to simplify $\frac{14}{24}$ to $\frac{7}{12}$."

"And $\frac{3,867}{8,978}$ of $\frac{7,324}{11,873}$ is the same as $\frac{3,867}{8,978} \times \frac{7,324}{11,873}$?" asked Barnaby.

"Um, yup," said Jack. "I'll let you do that one on your own."

"I've got one," said Bridget. "It's exactly $\frac{1}{2}$ a mile from here to my house. If I walk $\frac{1}{4}$ of the way to my house, how far will I have gone?"

"All you're asking is: What is $\frac{1}{4}$ of $\frac{1}{2}$? So we can set up the multiplication problem," said Beauregard.

$$\frac{1}{4} \times \frac{1}{2} =$$

"Just multiply straight across.

$$\frac{1}{4} \times \frac{1}{2} = \frac{1 \times 1}{4 \times 2} = \frac{1}{8}$$

and you would walk $\frac{1}{8}$ of a mile."

"How about this one?" Bridget continued. "Let's go back to Beauregard's catnip party."

"Please let me assure you that it was merely a polite gathering of old friends for stimulating conversation," said Beauregard, knowing that nobody would believe him.

"If Beauregard wanted to share the $\frac{1}{2}$ pound of catnip equally with his friends and he wanted to give each friend $\frac{1}{12}$ of a pound, how many shares can he make?"

"That's division!" said Barnaby. "You're asking how many times $\frac{1}{12}$ goes into $\frac{1}{2}$. That's the same as $\frac{1}{2} \div \frac{1}{12}$."

"To divide fractions," said Jack, "you invert the second fraction, or flip it over, and then multiply. So to do $\frac{1}{2} \div \frac{1}{12}$, you invert $\frac{1}{12}$ to get $\frac{12}{1}$, and then multiply $\frac{1}{2} \times \frac{12}{1}$. By the way, when you invert a fraction, mathematicians call that taking the **reciprocal**. So to divide fractions:

$$\frac{1}{2} \div \frac{1}{12} =$$

You invert the second fraction.

$$\frac{1}{12} \text{ becomes } \frac{12}{1}$$

Then multiply.

$$\frac{1}{2} \times \frac{12}{1} = \frac{12}{2} = \frac{6}{1} = 6$$

Beauregard can divide his catnip into 6 equal portions."

"You might have noticed from the catnip question that we got a fraction that was bigger than 1 ($\frac{12}{2}$ or $\frac{6}{1}$). That's okay, fractions can be as big as you want them to be. So for instance, $\frac{5}{2}$ is a perfectly good fraction, it just means that you have $\frac{2}{2} + \frac{2}{2} + \frac{1}{2}$, or $2\frac{1}{2}$. Probabilities can't be greater than 1, but only a small number of the fractions that you'll ever see will be probabilities," Jack said.

Bridget joined in, "Let's try another division question. After Barnaby destroyed $\frac{7}{12}$ of his laboratory, we set out to rebuild it. We found that we could only repair $\frac{1}{24}$ of the lab each day. How many days did it take us to repair the lab?

"So what we're asking is: How many times does $\frac{1}{24}$ go into $\frac{7}{12}$? Or, what is $\frac{7}{12}$ divided by $\frac{1}{24}$?

Write it down.

$$\frac{7}{12} \div \frac{1}{24} =$$

Invert the second fraction. That is, take the reciprocal of the second fraction.

$$\frac{1}{24} \text{ becomes } \frac{24}{1}$$

Now multiply.

$$\frac{7}{12} \times \frac{24}{1} = \frac{7 \times 24}{12 \times 1} =$$

Before we multiply, remember that we can make our lives easier by reducing the fraction. Let's divide the top and bottom of our fraction by 12. That will cancel out the 24 on top and the 12 on the bottom.

$$\frac{7 \times \overset{2}{\cancel{24}}}{\underset{1}{\cancel{12}} \times 1} = \frac{7 \times 2}{1 \times 1} = \frac{14}{1} = 14$$

So it took 14 days to repair Barnaby's lab."
Here are a few more questions for practice.

✎ ✎ ✎ ✎ ✎

12) Dr. Ismore likes to eat a little stinky cheese at night before he eats dinner. He bought $\frac{5}{6}$ of a wheel (stinky cheese is generally sold in circles called wheels) at the stinky cheese shop on his way home the other day. If he ate $\frac{1}{10}$ of what he bought, what fraction of a wheel did he eat?

13) Babette gave Dr. Ismore $\frac{1}{2}$ of a loaf of French bread to eat with his stinky cheese. If the professor ate $\frac{2}{11}$ of what Babette gave him, what fractional part of the entire loaf did he eat?

14) Barnaby has $\frac{3}{4}$ of a gallon of lemonade. How many glasses, each of which holds $\frac{1}{20}$ of a gallon, can he fill?

15) Bridget drank some of Barnaby's lemonade and got sick. She spent $\frac{2}{3}$ of a day in the emergency room. She was unconscious for

$\frac{1}{3}$ of the time that she was in the emergency room. What fraction of the day was she unconscious?

16) Barnaby has a mini-computer with 8 megabytes of memory. If memory for a mini-computer comes in segments that are each $\frac{1}{2}$ megabyte large, how many memory segments are in the computer?

✎ ✎ ✎ ✎ ✎

Once the gang had finished talking about fractions, they sat down to play some cards. Out of respect for the youth of our heroes, One-eyed Jack agreed to play for oatmeal cookies instead of money. Within an hour, Jack sat behind a mountain of cookies, while everyone else at the table sat nibbling on crumbs.

Jack apologized, saying, "I can't help it, my friends. When you're a professional, there's no such thing as a friendly game." As they all got up to go, he offered them each a cookie, as well as some parting advice: "Never draw to an inside straight, never play for money against a guy named Doc, and don't forget fractions; you always seem to need them when you least expect it."

👍 👍 👍 👍 👍 👍 👍 👍

ANSWERS:

1) There are a total of 6 around the table, so the whole is 6. There are 2 females, Bridget and Babette, so the female part is 2.

$$\frac{\text{part}}{\text{whole}} = \frac{2}{6}$$

We can simplify $\frac{2}{6}$ because the numerator and denominator are both even.

$$\frac{2 \div 2}{6 \div 2} = \frac{1}{3}$$

So $\frac{1}{3}$ of those around the table are female.

2) There are 12 "face cards" in a deck: 4 jacks, 4 queens, and 4 kings. So the fraction of the deck that is made up of face cards is 12. The whole deck contains 52 cards, so 52 is the whole.

$$\frac{\text{part}}{\text{whole}} = \frac{12}{52}$$

We can simplify this fraction by dividing both the numerator and denominator by 4.

$$\frac{12 \div 4}{52 \div 4} = \frac{3}{13}$$

So the probability of getting a face card from a shuffled deck of cards is $\frac{3}{13}$.

3) There were 3 dogs at the party, so the dog part is 3. The whole is the total number of animals, 5 + 3 + 1 = 9.

$$\frac{part}{whole} = \frac{3}{9}$$

Top and bottom are divisible by 3.

$$\frac{3 \div 3}{9 \div 3} = \frac{1}{3}$$

So $\frac{1}{3}$ of the animals at the party were dogs.

4) The whole is 8 gallons. The part is 2 gallons.

$$\frac{part}{whole} = \frac{2}{8} = \frac{1}{4}$$

So Barnaby used $\frac{1}{4}$ of the horse sweat in the recipe.

5) Use the bowtie.

$$\frac{5}{8} \diagup\!\!\!\!\diagdown \frac{7}{9}$$

Now multiply along the arrows and put the products above the fractions.

$$\overset{45}{\frac{5}{8}} \diagup\!\!\!\!\diagdown \overset{56}{\frac{7}{9}}$$

The bigger number is over the bigger fraction, so $\frac{7}{9}$ is bigger and Bridget has read the greater fraction.

6) Use the bowtie.

$$\frac{1}{3} \diagup\!\!\!\!\!\diagdown \frac{3}{10}$$

Now multiply along the arrows and put the products above the fractions.

$$\overset{10}{\frac{1}{3}} \diagup\!\!\!\!\!\diagdown \overset{9}{\frac{3}{10}}$$

The bigger number is over the bigger fraction, so $\frac{1}{3}$ is bigger and Barnaby has eaten more of the mashed potatoes.

7) Set up the problem.

$$\frac{1}{2} + \frac{3}{7} =$$

Multiply across the bottom to get a common denominator.

$$\frac{1}{2} + \frac{3}{7} = \frac{}{14}$$

Do the bowtie.

$$\overset{7}{\frac{1}{2}} \diagup\!\!\!\!\!\diagdown \overset{6}{\frac{3}{7}} = \frac{}{14}$$

Now add the numbers you got from the bowtie to get the numerator of your answer.

$$\frac{1}{2} + \frac{3}{7} = \frac{7 + 6}{14} = \frac{13}{14}$$

Barnaby's idea is $\frac{13}{14}$ baked. That's better than most of his ideas.

8) Set up the problem.

$$\frac{2}{5} + \frac{2}{9} =$$

Multiply across the bottom to get a common denominator.

$$\frac{2}{5} \longrightarrow \frac{2}{9} = \frac{}{45}$$

Do the bowtie.

$$\overset{18}{} \quad \frac{2}{5} \bowtie \frac{2}{9} \overset{10}{} = \frac{}{45}$$

Now add the numbers you got from the bowtie to get the numerator of your answer.

$$\frac{2}{5} + \frac{2}{9} = \frac{18 + 10}{45} = \frac{28}{45}$$

Babette and Bridget together have eaten $\frac{28}{45}$ of the pie, which is a little more than half.

9) This is subtraction. Set up the problem.

$$\frac{1}{4} - \frac{1}{6} =$$

Multiply across the bottom to get a common denominator.

$$\frac{1}{4} \longrightarrow \frac{1}{6} = \frac{}{24}$$

Do the bowtie.

$$\frac{\overset{6}{1}}{4} \longrightarrow \frac{\overset{4}{1}}{6} = \frac{}{24}$$

Now subtract the numbers you got from the bowtie.

$$\frac{1}{4} - \frac{1}{6} = \frac{6 - 4}{24} = \frac{2}{24}$$

We can reduce $\frac{2}{24}$.

$$\frac{2 \div 2}{24 \div 2} = \frac{1}{12}$$

The chances of drawing a spade from a deck of cards are greater than the chances of rolling a seven with a pair of dice by $\frac{1}{12}$.

10) Set up the addition problem.

$$\frac{2}{7} + \frac{1}{5} =$$

Multiply across the bottom to get a common denominator.

$$\frac{\overset{6}{1}}{4} \longrightarrow \frac{\overset{4}{1}}{6} = \frac{}{24}$$

Do the bowtie.

$$\overset{10}{\frac{2}{7}} \times \overset{7}{\frac{1}{5}} = \frac{}{35}$$

$$\frac{2}{7} + \frac{1}{5} = \frac{10 + 7}{35} = \frac{17}{35}$$

Lefty and Righty have won $\frac{17}{35}$ of the Yankees victories. That's just a bit less than half. (Check it using the bowtie if you want.)

11) Set up the subtraction problem.

(Mouthwash initially in the bottle) – (Mouthwash left at the end) = (Mouthwash Dr. Ismore drank)

$$\frac{3}{4} - \frac{1}{5} =$$

Multiply across the bottom to get a common denominator.

$$\frac{3}{4} \longrightarrow \frac{1}{5} = \frac{}{20}$$

Do the bowtie.

$$\overset{15}{\underset{4}{\frac{3}{4}}} \times \overset{4}{\underset{}{\frac{1}{5}}} = \frac{}{20}$$

Now subtract the numbers you got from the bowtie.

$$\frac{3}{4} - \frac{1}{5} = \frac{15 - 4}{20} = \frac{11}{20}$$

Dr. Ismore drank $\frac{11}{20}$ of the bottle, more than half. He claimed it was actually very tasty and it left his breath minty fresh.

12) The question is: What is $\frac{1}{10}$ of $\frac{5}{6}$? So we set up the multiplication problem.

$$\frac{1}{10} \times \frac{5}{6} =$$

Multiply straight across the top and bottom.

$$\frac{1}{10} \times \frac{5}{6} = \frac{1 \times 5}{10 \times 6} = \frac{5}{60},$$ which reduces to $\frac{1}{12}$.

Dr. Ismore ate $\frac{1}{12}$ of a wheel of stinky cheese. Maybe he washed it down with the mouthwash.

13) The question is: What is $\frac{2}{11}$ of $\frac{1}{2}$? More multiplication.

$$\frac{2}{11} \times \frac{1}{2} =$$

Multiply the numerators and denominators.

$$\frac{2}{11} \times \frac{1}{2} = \frac{2 \times 1}{11 \times 2} = \frac{2}{22},$$ which reduces to $\frac{1}{11}$.

14) Here's a division problem. How many times does $\frac{1}{20}$ go into $\frac{3}{4}$? That is, what is $\frac{3}{4}$ divided by $\frac{1}{20}$?

$$\frac{3}{4} \div \frac{1}{20} =$$

Invert the second fraction.

$$\frac{1}{20} \text{ becomes } \frac{20}{1}$$

Now multiply.

$$\frac{3}{4} \times \frac{20}{1} = \frac{60}{4}, \text{ which reduces to 12.}$$

Barnaby can fill exactly 12 glasses of lemonade.

15) Multiplication. What is $\frac{1}{3}$ of $\frac{2}{3}$? Just multiply straight across the top and bottom.

$$\frac{1}{3} \times \frac{2}{3} = \frac{2}{9}$$

Bridget was unconscious for $\frac{2}{9}$ of the day. Barnaby felt terrible, but Bridget felt worse.

16) You don't really need to know anything about computers to answer this division problem. All they are asking is how many times $\frac{1}{2}$ goes into 8. That's the same as $\frac{8}{1}$ divided by $\frac{1}{2}$.

$$\frac{8}{1} \div \frac{1}{2} =$$

Invert the second fraction.

$$\frac{1}{2} \text{ becomes } \frac{2}{1}$$

Now multiply.

$$\frac{8}{1} \times \frac{2}{1} = \frac{16}{1} = 16$$

The computer has 16 memory segments.

Chapter 5
Decimals: Hanging Out at the DeciMall

"There's one thing I don't get," said Barnaby when they had returned to the professor's study. "When we were playing cards with One-Eyed Jack, you said that a fraction was the same thing as division. Remember we said:

$$\frac{\text{anything}}{\text{anything else}} = \text{anything} \div \text{anything else.}$$

"Now, that's okay if the numerator and denominator work out nicely and evenly, like $\frac{8}{2}$, or $\frac{9}{3}$, or $\frac{20}{5}$, but what do we do if the top number is smaller than the bottom number, like $\frac{2}{5}$, or if the denominator doesn't go into the numerator evenly, like $\frac{10}{4}$?"

"I think it's time for a trip to the DeciMall," said Dr. Ismore.

"The deciwhat?" asked Bridget.

"The DeciMall," repeated the professor. "Some numbers are best written as **decimals**, and the best place to look for decimals is at the DeciMall. Babette, please run down to the mailbox and bring back a mail-order catalog."

Babette ran downstairs and returned after a few moments with an armload of catalogs. "The mailman just left these. Poor man, he looked like his back was breaking. There are so many, which one do you want?"

"It doesn't matter," said the professor as he took the top catalog off the pile. He opened the catalog, which seemed to offer a combination of fruit, computer software, and underwear, and began to chant an incantation in a low mumble, "Lowpricesgreatselectionfreeparkinglow-pricesgreatselectionfreeparking." In a few seconds, they were seated in the food court at the DeciMall.

"If DeciMall food is anything like shopping mall food, I suggest we skip lunch and go straight to the information booth," said Beauregard, and everybody agreed.

The information booth was at the center of the mall and in the information booth was a kindly looking information lady. "Welcome," she said. "I'm Ms. LaPoint, how can I help you?"

The professor answered, "We were wondering if you could point us toward the best place to find out about decimals."

"I know just the point where you should start," Ms. LaPoint pointed out.

"I think I'm getting the point," said Bridget.

The gang gathered around Ms. LaPoint as she began her explanation. "The thing that makes a decimal a decimal is the decimal point. The decimal point separates the part of the number that's bigger than 1 from the part that's smaller than 1. It sits to the right of the ones digit in any number.

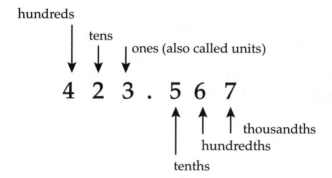

So the number 423.567 is a compact way of saying:

$$400 + 20 + 3 + \frac{5}{10} + \frac{6}{100} + \frac{7}{1,000}$$

And 5.870 is

$$5 + \frac{8}{10} + \frac{7}{100} + \frac{0}{1,000}$$

"Any number can be written with a decimal point, but we don't usually bother with the decimal point for whole numbers. But if we wanted, we could write 23 as 23, or 23.0, or even 23.000000. Adding zeroes on the far right of a decimal number doesn't change the value of the number, but sometimes adding zeroes is a big help. In fact, that's the key to answering the question Barnaby asked about dividing numbers that don't divide nicely. Let's try $\frac{10}{4}$.

First set up the long division.

$$4\overline{)10}^{\,?}$$

Now do what you can.

$$
\begin{array}{r}
2 \\
4{\overline{)10}} \\
\underline{8} \\
2
\end{array}
$$

We could just say that our answer is 2 r.2 (that's 2 with a remainder of 2), but we can do better than that. Let's add a decimal point and a zero to the 10. Remember, adding a decimal point and zeroes after a whole number doesn't change the value of the number.

$$
\begin{array}{r}
2.5 \\
4{\overline{)10.0}} \\
\underline{8} \\
2\,0 \\
\underline{2\,0} \\
0
\end{array}
$$

All you need to remember is to put the decimal point in your answer directly above the decimal point under the division roof.

"Let's try $\frac{2}{5}$."
"Set it up."

$$5\overline{)2}^{?}$$

"Since 5 doesn't go into 2, you can't do anything until you've put in the decimal point and a zero."

$$
\begin{array}{r}
0.4 \\
5\overline{)2.0} \\
2\,0 \\
\hline
0
\end{array}
$$

"So now you can divide any two numbers, even if they don't divide evenly. You can also convert any fraction to a decimal," Ms. La Point finished.

Now you try some.

1) Dr. Ismore bought 4 pretzels for the group. He would have bought 5, but cats don't like pretzels. If he paid $5 total for the pretzels, how much did each one cost?

2) Babette punches 6 ÷ 5 = into a calculator. What is the answer she sees on the digital readout?

3) How would the fraction $\frac{1}{8}$ be written as a decimal?

"What do I do if the numbers I'm dividing already have decimal points?" asked Beauregard. "What if mouse brains cost $.50 per pound and I have $3.25. How many pounds of mouse brains can I afford?"

"Mouse brains?!" said Babette.

"They're too small for a main course," said Beauregard, "but they make excellent appetizers if you serve them with crackers."

"Is anybody else getting hungry?" asked the professor.

"First let's answer my question," said Beauregard.

Ms. LaPoint continued, "The question is: What is 3.25 ÷ .50? First, remember how we could add as many zeroes on the far right side of the decimal as we wanted without changing the value of the number? Well we can also get rid of zeroes if we don't want them. So let's write .50 as .5. That doesn't change the value, but it makes it simpler to deal with. Now set up the division.

$$.5\overline{)3.25}^{?}$$

In order to divide decimals, you first have to make the number on the left (that's .5 here) into a whole number. You do that by moving the decimal point to the right. In this case you move it one place to the right, turning .5 into 5. But anything you do to the number on the left, you also have to do to the number on the right, so you have to turn 3.25 into 32.5. Like this:

$$.5\underset{\curvearrowright}{)}\,3.2\underset{\curvearrowright}{5} \implies 5.\overline{)\,32.5}$$

Now you can divide like we did before. Just remember to move the decimal point straight up to your answer on the division roof.

$$5\overline{)\,32\overset{\cdot}{.}5}$$

Let's do the math.

$$5 \overline{)32.5} \quad \begin{array}{r} 6.5 \\ \hline \end{array}$$

```
     6.5
5)32.5
  30
  ---
   2 5
   2 5
   ---
     0
```

Beauregard can afford 6.5 pounds of mouse brains. Here's another. If Barnaby plays Deathpong, a video game that costs $.25 per play, and he spends $4, how many games does he play? First set up the division.

```
      ?
.25)4
```

Now move the decimal point.

.2̦5̦)̄ 4̦. ⌣⌣ ⟹ 25)̄ 400

Remember that 4 is the same as 4.00, so when we move the decimal point two places to the right, we get 400. Now divide.

```
     16
25)400
   25
   ---
   150
   150
   ---
     0
```

Barnaby gets 16 plays for his four bucks."
Hey reader, why don't you try a few more?

✎ ✎ ✎ ✎ ✎

4) Beauregard has 3.6 ounces of catnip and he wants to give some to his friends. If he has bags that hold 0.6 ounces each, how many bags can he fill?

5) Write $\dfrac{0.3}{1.2}$ as a single decimal number.

6) Beauregard paid $6.82 for 3.1 pounds of mouseloaf. He got the expensive kind, with hair and eyes included. How much did he pay per pound?

✎ ✎ ✎ ✎ ✎

The gang's travels around the DeciMall eventually and inevitably led them to the candy counter. A large sign by the cash register advertised a special sale on jelly peas.

"What are jelly peas?" Bridget asked the clerk.

"The jelly bean shaper wasn't working yesterday, so today we have jelly peas," the clerk replied. "But they taste almost as good, and they're only $1.25 per pound. Last week we had jelly kumquats; we could barely *give* those away."

"I wonder how much 3.8 pounds of jelly peas would cost?" wondered the professor.

"I was wondering when we'd get around to multiplying decimals," wondered Beauregard.

"Actually, multiplying decimals is pretty easy when you compare it to dividing decimals. We've already done all the hard stuff for decimals," Bridget pointed out. "Aren't you supposed to do the easy stuff first and work up to the hard stuff?"

"Normally, yes," said the professor, "but I wanted to do division while you had some energy. To figure out how much the jelly peas cost, you need to multiply the price per pound, $1.25, by the number of pounds, 3.8. To multiply decimals, you set up the multiplication exactly the same way that you do for whole numbers.

$$\begin{array}{r} 1.25 \\ \times\, 3.8 \\ \hline \end{array}$$

"Now count the number of places to the right of the decimal points. In this case there are three."

$$
\begin{array}{r}
\overset{①\,②}{1.\underline{25}} \\
\times\ 3.\underline{8}\ ③ \\
\hline
\end{array}
$$

"Now multiply as if the decimal points weren't there."

$$
\begin{array}{r}
1.25 \\
\times\,3.8 \\
\hline
1000 \\
3700 \\
\hline
4700 \\
\end{array}
$$

"Finally, count three places from the right of the answer and put the decimal point back in."

$$
4\underset{3}{.}\underset{2}{7}\underset{1}{5}\,0
$$

"So 3.8 pounds of jelly peas cost $4.75."

Bridget saw another sign on the candy counter. "What are lemons dropped?" she asked. "Shouldn't that say *'lemon drops'*?"

"No, those are the ones that fell on the floor in the kitchen," answered the clerk. "They're only $.40 a pound and they're fine as long as you wipe them off before you eat them."

"I'll take a few," said Bridget. "Give me .05 pounds." She stepped over to Barnaby, who still wasn't paying any attention. "Here you go, Barnaby," she said. "I got you some lemon drops. Enjoy."

Let's figure out how much she paid for them. First set up the multiplication.

$$
\begin{array}{r}
.40 \\
\times\,.05 \\
\hline
\end{array}
$$

Count the number of places to the right of the decimal points. Here, all of the places are to the right of the decimal points, for a total of four. Now ignore the decimal points and multiply.

$$\begin{array}{r} .40 \\ \times\ .05 \\ \hline 200 \end{array}$$

Now count four decimal places from the right. In this case we have to add zeroes to the left of the number.

The answer is .0200. Bridget paid $.02, or 2 cents, for her lemons dropped. Now you try some problems.

7) Each of Barnaby's fingers is 3.1 inches long and he's got 10 fingers. What is the total length of his fingers?

8) At a perfume shop at the DeciMall, Babette's favorite perfume, "Eau du Fromage," costs $6.90 per ounce. How much would 1.7 ounces cost?

9) Beauregard bought a ball of yarn as a birthday gift for a kitten. If yarn costs $.04 per foot and the ball contains 30.25 feet of yarn, how much was the ball of yarn?

"What about addition," asked Babette. "What if I've spent $1.28 on jelly peas and $1.79 on lemons dropped. How much did I spend?"

"Adding decimals is easy," said Bridget. "To add decimals, all you have to do is stack up the numbers, lining up the decimal points like this:

$$\begin{array}{r} 1.28 \\ + 1.79 \\ \hline 3.07 \end{array}$$

Once you line up the decimal points, you can add decimals just like you add whole numbers. So your candy cost you $3.07."

"And subtraction?" Babette continued. "What if I gave the man at the counter a $10 bill. How much change did I receive?"

"Subtraction is like addition," Bridget explained. "You have to line up the decimal points. Remember that you can put the decimal point at the end of any whole number. So you can set it up like this:

$$\begin{array}{r} 10. \\ - 3.07 \\ \hline \end{array}$$

That still looks kind of ugly, but you can always add zeroes at the right of a decimal number without changing its value. If we add some zeroes after the 10, the problem will look better:

$$\begin{array}{r} 10.00 \\ - 3.07 \\ \hline \end{array}$$

Now we can subtract.

$$\begin{array}{r} 10.00 \\ - 3.07 \\ \hline 6.93 \end{array}$$

So, Babette, your change should have been $6.93."

"That's the way I like to add and subtract decimals," Barnaby added. "First I line up the decimals, then I add zeroes on the right until both of the numbers line up."

Bridget agreed. "So if I want to add 45.6 to 2.352, the first thing I do is line up the decimal points.

$$45.6$$
$$+\ 2.352$$

Now I add zeroes on the right to make the addition easier to deal with,

$$45.600$$
$$+\ 2.352$$

and all that's left to do is add."

$$45.600$$
$$+\ 2.352$$
$$\overline{47.952}$$

Here's some more practice stuff. (The answers are on page 85.)

10) Bridget is 4.2 feet tall and Babette is 5.123 feet tall, what is their combined height?

11) If you had $23.76 in your pocket and you spent a $5 bill, how much would you have left?

12) Before they left for the DeciMall, Dr. Ismore weighed 195.1 pounds. After the trip to the candy counter, the professor found that he had gained .26 pounds. How much did the professor weigh after the trip?

"Has anybody seen Beauregard?" asked Barnaby.

"I saw him wander off that way a little while ago," said Bridget, pointing to her right. "Over there, by . . . the uh-oh!" Bridget broke into a run without finishing her sentence, heading in the direction that she had pointed.

"Where are we going?" Babette shouted breathlessly as she and the group tried to catch up with Bridget.

"The fountain!" Bridget yelled over her shoulder, pointing to a large fountain in the center of the mall.

They were a few steps too late. As they approached the fountain, they heard a loud splash, followed by the gurgles and yowls of a normally immaculately groomed cat who hated being wet more than anything in the world, but loved the taste of fountain-fattened goldfish almost as much as he hated being wet. Beauregard, it seemed, had lost his footing while taking a stab at an especially plump fish that had moved out of the way deceptively quickly for her size.

Without hesitating, Dr. Ismore leapt off the edge of the fountain in a gracefully executed swan dive, swam out to the center, and retrieved Beauregard using perfect Red Cross lifesaving technique. As the professor swam back to the edge of the fountain and deposited the wet cat on dry land, the gang marvelled that he had managed a textbook example of a water rescue in a pool that was only a foot and a half deep.

Beauregard was unhurt, except for his dignity. There's nothing less dignified than a wet cat, unless of course its a waterlogged, fully clothed, cue-ball bald professor who is just realizing that a fountain is only a foot and a half deep and is trying to stand up on the slippery bottom.

"Now I'm in trouble," said Dr. Ismore as he sat in the water and watched the mall security guards arrive and cautiously circle the fountain. "Another scandal this year and I'll probably lose my job. And I promised the dean that I wouldn't get caught swimming in mall fountains again."

"Again!?" Babette cried.

"Yes, it's a long story, and not very pretty," said the

professor with a sigh and a gurgle. "I wish there was some way we could get out of here before anyone reports this."

"I've got it!" said Barnaby. "Quick, professor, give me the mail-order catalog." Dr. Ismore reached into his coat pocket and handed Barnaby the sopping-wet catalog. Barnaby opened the catalog and quickly turned the pages until he found the home delivery section.

"Rushhomedeliveryavailableonrequest," muttered Barnaby, and in a moment they were back in the professor's study. Beauregard and the professor stood dripping in the middle of the room.

"Let me get you some towels; you're dripping all over the place," said Babette.

"Don't worry, it's good for the carpet," said Dr. Ismore. "Good thinking, Barnaby, I'd forgotten about the catalog."

"You'd think they could clean the water in that fountain once in a while," Beauregard sniffed as he tried to return his fur to its normal tidy state.

"Perhaps you'll remember this the next time you want to go fishing in a fountain," said Babette sternly.

Beauregard emitted a polite little cat belch that had a pleasant goldfish aftertaste. "Well, some things are worth trading a little dignity for," he said with a sly smile.

ANSWERS:

1) The question is: What is $5 \div 4$? (That's the same thing as converting $\frac{5}{4}$ to a decimal.) We'll need two zeroes after the decimal point for this one.

$$
\begin{array}{r}
1.25 \\
4\overline{)5.00} \\
\underline{4} \\
10 \\
\underline{8} \\
20 \\
\underline{20} \\
0
\end{array}
$$

Each pretzel costs $1.25.

2) What is $6 \div 5$?

$$
\begin{array}{r}
1.2 \\
5\overline{)6.0} \\
\underline{5} \\
1\,0 \\
\underline{1\,0} \\
0
\end{array}
$$

The calculator says 1.2. You'll be spending a lot of time working with calculators in your math career. That's one of the reasons why it's so important to know how to work with decimals.

3) What is $1 \div 8$?

$$
\begin{array}{r}
0.125 \\
8\overline{)1.000} \\
\underline{8} \\
20 \\
\underline{16} \\
40 \\
\underline{40} \\
0
\end{array}
$$

So $\dfrac{1}{8} = 0.125$.

4) What is $3.6 \div .6$? First set it up.

$$
\begin{array}{r}
? \\
.6\overline{)3.6}
\end{array}
$$

Move both decimal points one place to the right and divide.

$$
\begin{array}{r}
6 \\
6.\overline{)36.} \\
\underline{36} \\
0
\end{array}
$$

Beauregard can fill 6 bags of catnip.

5) $\dfrac{0.3}{1.2}$ is the same as $0.3 \div 1.2$. Set it up.

$$
\begin{array}{r}
? \\
1.2\overline{).3}
\end{array}
$$

Move both decimal points one place to the right and divide. Remember to move the decimal point straight up to your answer on the roof of the division sign. You'll need to add two zeroes on the right for this one.

$$
\begin{array}{r}
.25 \\
12\overline{)3.00} \\
2\,4 \\
\hline
60 \\
60 \\
\hline
0
\end{array}
$$

6) What is 6.82 ÷ 3.1? Set it up.

$$
3.1\overline{)6.82}^{\;?}
$$

Move both decimal points one place to the right and divide.

$$
\begin{array}{r}
2.2 \\
31\overline{)68.2} \\
62 \\
\hline
6\,2 \\
6\,2 \\
\hline
0
\end{array}
$$

Beauregard pays 2.2 dollars (that's $2.20) per pound of mouseloaf with hair and eyes. Without hair and eyes, it's only $1.50 per pound, but everybody knows that all the flavor is in the hair and eyes.

7) Set up the multiplication and count the decimal places. This time there's only one place to the right of the decimal point.

$$
\begin{array}{r}
3.1 \\
\times\ 10 \\
\hline
\end{array}
$$

Multiply.

$$
\begin{array}{r}
3.1 \\
\times\ 10 \\
\hline
310 \\
\end{array}
$$

Move the decimal point one space to the left to get 31 as the answer. Barnaby's fingers are a total of 31 inches long. Did you notice that when you multiplied 3.1 by 10, you got 31, which is the same number with the decimal point in a different place, one place to the right? That always happens. When you multiply a number by 10, it's the same as just moving the decimal point one place to the right.

8) Set up the multiplication and count the decimal places. There are three here.

$$
\begin{array}{r}
6.90 \\
\times\ 1.7 \\
\hline
\end{array}
$$

Now ignore the decimal points and multiply.

$$
\begin{array}{r}
6.90 \\
\times\ 1.7 \\
\hline
4830 \\
6900 \\
\hline
11730 \\
\end{array}
$$

Move the decimal point three places from the right and 11730 becomes 11.730. In dollars and cents, Eau du Fromage costs $11.73.

9) Set it up and count the decimal places. Here there are four.

$$\begin{array}{r} 30.25 \\ \times\ \ .04 \\ \hline \end{array}$$

Ignore the decimal points and multiply.

$$\begin{array}{r} 30.25 \\ \times\ \ .04 \\ \hline 12100 \end{array}$$

Move the decimal point four places from the right and 12100 becomes 1.2100. Remember, we can drop the zeroes on the far right of the decimal point to get 1.21. So the yarn costs Beauregard $1.21.

10) Line up the decimal points, add some zeroes on the right, and add.

$$\begin{array}{r} 4.200 \\ +\ 5.123 \\ \hline 9.323 \end{array}$$

Bridget and Babette's combined height is 9.323 feet tall.

11) Line up the decimal points, add some zeroes on the right, and subtract.

$$\begin{array}{r} 23.76 \\ -\ 5.00 \\ \hline 18.76 \end{array}$$

You've got $18.76 left.

12) Line up the decimal points, add a zero on the right, and add.

$$
\begin{array}{r}
195.10 \\
+ \quad .26 \\
\hline
195.36
\end{array}
$$

Dr. Ismore weighs 195.36 pounds.

Chapter 6
Percents: Would You Buy a Used Car From This Woman?

Professor Les Ismore leaned back in his chair and gazed thoughtfully at the group assembled in his study. He coughed in that funny way that made it clear that he didn't really have to cough; he just wanted everybody's attention. He looked like a man with something important to say, and this was not lost on Bridget, Barnaby, and Babette, who put down the books they were reading and gave the professor their attention. Beauregard was napping in a comfortable chair, but he was able to raise his ears in such a way that made people think he was paying attention, and that's what he did.

"I've had a pretty good life," the professor began. "I've written a hundred books and invented a new branch of mathematics; I've won the Nobel and Pulitzer prizes; I have a lovely family, wonderful friends, worthy associates, and brilliant students, but still I feel that I'm missing something very important."

"You could start your own university," suggested Barnaby.

"No, that's not it," the professor sighed.

"Maybe you could enter politics," Bridget tried. "Is it power that you're after?"

"No, it goes much deeper than that," said the professor.

"What is it, then?" asked Babette. "Is it something you can put into words?"

"Sure," said the professor. "I need wheels."

"Wheels?!" said the gang in unison.

"Yup. Wheels. A guy's life just isn't complete if he doesn't have a really cool car," Dr. Ismore explained. "I've been searching for the right one for a long time, but I still haven't found exactly what I'm looking for. I think it's time we went car shopping."

Everyone agreed that car shopping was a fine idea. Dr. Ismore told Barnaby to go over to the shelf by the window and get the Staten Island phone book. He then sent Bridget upstairs to retrieve the owner's manual for the 1977 AMC Pacer. Everyone gathered around the professor as he set the two books on the table. He opened the owner's manual to the emergency procedures page, and he opened the phone book to the used car dealerships section. Then he told everyone to blink at once.

They opened their eyes in a haze of cigar smoke. Through the haze was a plate glass window that looked out at the sun-drenched parking lot of a used car dealership. They sat in uncomfortable plastic chairs in the office of the dealership and faced the owner, who sat at a simple metal desk. Beauregard coughed.

"If that cat brings up a hairball in my office, I'm going to be one unhappy camper dealer," said the owner, a block-shaped woman in a three-piece suit who formed her words around the cigar she kept clenched between her teeth.

"Relax, Marge, it's the smoke," said the professor.

"Ah, of course," Marge replied, but it didn't occur to her to put out the cigar. "Welcome back to Staten Island, Les. You always seem to just appear in my office out of nowhere."

"It's cheaper than taking the subway and then the ferry," said the professor. Dr. Ismore introduced Marge to the gang and explained that he was still in the market for his dream car.

Marge got up from the desk and led them out to the lot. "Dream cars, muscle cars, sports cars, trucks, vans, campers, I've got them all. How about you?" Marge said, gesturing toward Babette. "You look like the motorcycle type. I've got a cherry red, American-made bike. One previous owner: a little old lady who kept it in a heated garage and only took it out once a year to drive to Sturgis, South Dakota, and back. All highway miles. I got it for 40 percent of what it was worth and I marked it up 10 percent from there. But I like you and you're a friend of Dr. Ismore's, so I'll knock 20 percent off the current price." She turned to the professor. "How about a pickup truck? I've got a beauty for you. The previous owner was a rodeo clown who only used it to take bull riders to the hospital. Now, normally I discount almost everything on the lot 20 percent, but you're a friend of Babette's so I'll take off another 15 percent. You'd be paying less than 50 percent of the blue book value." Marge then stopped to explain that the blue book is the book used car dealers use to tell what a car is supposed to be worth.

"Well, first how about explaining **percents** to us?" asked the professor.

Marge grinned. "It would be my pleasure."

Marge explained that percents are really just fractions. A percent is just a specific kind of fraction in which the bottom number (the denominator, remember?) is always 100. So 50% is just $\frac{50}{100}$, and 15% is the same as $\frac{15}{100}$.

"That's all there is to it? You mean 84% is just $\frac{84}{100}$?" asked Bridget.

"That's it. Percents and fractions are the same thing," said Marge. "So let's say that motorcycle I got was worth $800. If I paid 40% of what it was worth, how much did I pay?"

Barnaby spoke up. "So you're asking: What is 40% of $800?"

"Exactly," said Marge.

"Remember when we talked about fractions," said Babette. "We found out that when you see the word 'of' it means you have to multiply. And if 40% is just $\frac{40}{100}$, then 40% of 800 is the same as $\frac{40}{100} \times \frac{800}{1}$ and all we're doing is multiplying fractions.

Let's do the math.

$$40\% \text{ of } 800 = \frac{40}{100} \times \frac{800}{1} = \frac{40 \times 800}{100 \times 1} =$$

Remember that when you multiply fractions, you can multiply straight across the top and bottom. You can also cancel things that are on the top and bottom to make your life easier. Let's cancel.

$$\frac{40 \times \overset{8}{\cancel{800}}}{\underset{1}{\cancel{100}} \times 1} \;=\; \frac{40 \times 8}{1} \;=\; 320$$

We've found out that 320 is 40% of 800. So Marge paid $320 for the motorcycle.

"Here's another one. Let's say Marge offered the professor a special deal on the pickup truck. The price of the truck was $4,400, but Marge said that he could drive the truck away right then if he gave her 25% that day and the rest in low monthly payments. How much would the professor have to give Marge to drive the truck away?

The question is: What is 25% of $4,400?

$$25\% \text{ of } 4,400 = \frac{25}{100} \times \frac{4,400}{1} = \frac{25 \times 4,400}{100 \times 1}$$

Let's cancel.

$$\frac{\overset{1}{\cancel{25}} \times 4,400}{\underset{}{\cancel{100}} \times 1} \;=\; \frac{4,400}{4} \;=\; 1,100$$

So Dr. Ismore can get himself a pickup truck for a $1,100 down payment."

In that last example, we reduced 25% to $\frac{1}{4}$. There are a few other percents that reduce to pretty simple fractions and you should memorize them. Check out the chart below:

PERCENT	FRACTION
50 %	$\frac{1}{2}$
$33\frac{1}{3}$ %	$\frac{1}{3}$
25%	$\frac{1}{4}$
20%	$\frac{1}{5}$
10%	$\frac{1}{10}$

Here are some problems for practice. (The answers are on page 102.)

1) At the beginning of the day, Marge had 45 cars on her lot. By the end of the day, she had sold 20% of the cars on the lot. How many cars did she sell?

2) Beauregard and his friends had dinner at a restaurant. The waitress gave them a very good table despite the strict "no furry animals" policy at the door. She had the chef make mousaroni and cheese for Beauregard, even though it wasn't on the menu. The cost of dinner was $160 and

Beauregard gave the waitress a generous 25% tip. How much did Beauregard tip the waitress?

3) Bridget went to 35 Yankees games last year. If the Yankees won 60% of those games, how many of the games that Bridget attended did the Yankees win?

4) The Statue of Liberty was a gift from France to America in 1884, and Babette insisted on visiting it on the way back from Staten Island. Babette climbed 85% of the way to the top of the statue. If the statue rises 300 feet above the ground, how high did Babette climb?

✎ ✎ ✎ ✎ ✎

Marge continued her sales pitch. "Les, how about a cargo van? They're durable and they've got plenty of room. I can give you a great deal on this red one over here. I bought it from a punk-rock band when they broke up after their record stiffed. It runs great, but I've got to let it go cheap because we've never been able to get the smell out of it."

The professor shook his head. "Sounds tempting, but I don't think so. What I'd really like is to hear more about percents. You've told us how to take a percent of a number. Now I'd like to know how to get a percent if I already have the numbers. Let's say you've got 50 cars on this lot and I've looked at 40 of them. What percent of the cars have I looked at?"

"No sweat," said Marge. "After all, we already know that a percent is just a fraction. So if we can make a fraction, we can make a percent."

Marge reminded everyone that a fraction is just $\frac{\text{part}}{\text{whole}}$. The part of the lot that the professor had seen was 40 cars, and the whole lot was 50 cars, so $\frac{\text{part}}{\text{whole}} = \frac{40}{50}$.

"Now comes the cool part," said Marge. "To convert

a fraction to a percent, all you have to do is multiply by 100."

"That's it?" asked the professor.

"That's it," Marge assured him. "Like this:"

$$\text{percent} = \text{fraction} \times 100 = \frac{\text{part}}{\text{whole}} \times \frac{100}{1}$$

Let's figure out what percent of the lot Dr. Ismore has seen.

$$\text{percent} = \frac{\text{part}}{\text{whole}} \times \frac{100}{1} = \frac{40}{50} \times \frac{100}{1} = \frac{40 \times 100}{50 \times 1} =$$

Now reduce before you multiply.

$$\frac{\overset{4}{\cancel{40}} \times 100}{\underset{5}{\cancel{50}} \times 1} = \frac{4 \times \overset{20}{\cancel{100}}}{\underset{1}{\cancel{5}} \times 1} = \frac{4 \times 20}{1} = 80$$

The professor has seen 80% of the cars on the lot.

"That's not bad," said Beauregard. "If I have a fraction, all I have to do is multiply it by 100 to convert it into a percent."

"Let's try another," said Marge. "There are six of us here, and three of us are female. What percent of this group is female?

Well, first we need to get a fraction. The female part of the group is 3 and the whole group is 6, so $\frac{\text{part}}{\text{whole}}$ is $\frac{3}{6}$.

$$\text{percent} = \frac{\text{part}}{\text{whole}} \times \frac{100}{1} = \frac{3}{6} \times \frac{100}{1} = \frac{3 \times 100}{6 \times 1}$$

Reduce before you multiply.

$$\frac{\overset{1}{\cancel{3}} \times 100}{\underset{2}{\cancel{6}} \times 1} = \frac{100}{2} = 50$$

So 50% of the group at the car dealership is female. Maybe you recognized right away that $\frac{3}{6}$ was the same as $\frac{1}{2}$, and you had already memorized $\frac{1}{2}$ = 50%. In that case, you didn't have to do any work.

Try some more (The answers are on page 103.):

✐ ✐ ✐ ✐ ✐

5) The gang has just gotten a new recording by the hot, new blues star, Blind Boneless Chicken. There are 15 songs on the disc and so far they've listened to 12 of them. What percent of the songs have they heard so far?

6) Marge bought a new cigar that was 8 inches long. She lit it and smoked it until a choking, gasping, coughing fit forced her to put it out. If the cigar was 6 inches long when she put it out, what percent of the cigar was left?

7) If you don't have a Staten Island phone book and an owner's manual for the 1977 AMC Pacer, you need to take the train and the ferry to get to Staten Island. If the trip takes 20 minutes on the train and 40 minutes on the ferry, what percent of the trip time is spent on the train?

8) Beauregard has prepared a big pot of mouse guts goulash. If he serves $\frac{7}{20}$ of the gou-

lash he's made with a little butter and some egg noodles, what percent of the goulash has he served?

✎ ✎ ✎ ✎ ✎

"How about that car over there? What's its story?" asked the professor.

Marge looked across the lot. "Tie-dyed, 1966 micro-bus. I bought it from a bunch of hippies when they converted their commune into an investment banking service. They did a lot of work on the engine. Now it runs on a mixture of yogurt, banana peels, and bran flakes. It doesn't get such great mileage, but the exhaust smoke smells like a banana cream pie."

The professor shook his head. "Not that one, the one behind it."

"Ah, yes," said Marge. "I knew you'd spot it eventually. It's the best deal on the lot; one of a kind, really." Marge described the car's history as they walked across the lot towards a vehicle that really was unlike anything they'd ever seen before. "I got it from an automobile-factory worker who took a single car part home with him from work every day for twenty years. He collected parts one piece at a time and kept them in his garage until he had all the pieces he needed. Then he built this car." They looked at the car, which seemed to relate the entire history of auto manu-facturing in one, four-wheeled hunk of steel. It had two headlights on the left side and only one on the right, but it had two doors on the right and only one on the left. In the rear, it had a trunk *and* a hatchback, but only one tail fin. No two pieces on the entire car were the same color. There were six windshield wipers, but only three of them were mounted on the windows.

"How much?" asked Dr. Ismore.

Marge started to negotiate. "I was hoping to get $5,500, but I did promise you a 20% discount."

"Before we go on, why don't you tell us about percent increase and decrease?" asked the professor.

"It would be my pleasure," said Marge.

Marge explained how to get percent increase or decrease from a number. For instance, here's how to get the new price of Dr. Ismore's car when it's decreased by 20% from $5,500.

First take 20% of $5,500.

$$\frac{20}{100} \times \frac{5,500}{1} = \frac{1}{5} \times \frac{5,500}{1} = \frac{1 \times 5,500}{5 \times 1} =$$

$$\frac{5,500}{5} = 1,100$$

The original price has been decreased by $1,100, so all we have to do is subtract 1,100 from 5,500.

$$\begin{array}{r} 5,500 \\ -1,100 \\ \hline 4,400 \end{array}$$

The new price is $4,400. So to find a 20% decrease from a number, all you have to do is figure out 20% of the original number and subtract it from the number.

Percent increase is calculated in almost the same way. Let's say that the professor agrees to buy the car for $4,400, but then he has to pay a 10% sales tax. How much does he actually pay for it?

First find 10% of $4,400.

$$\frac{10}{100} \times \frac{4,400}{1} = \frac{1}{10} \times \frac{4,400}{1} = \frac{1 \times 4,400}{10 \times 1} = \frac{4,400}{10} = 400$$

Now we're calculating percent increase, so we *add* 440 to 4,400.

$$
\begin{array}{r}
4,400 \\
+440 \\
\hline
4,840
\end{array}
$$

So Dr. Ismore pays $4,840, after tax, for the car. To find a 10% increase from a number, you take 10% of the number and add it to the number. Let's try a few. (The answers are on page 104.)

9) Let's go back to Beauregard and his dinner. If you recall, Beauregard gave a 25% tip after a $160 dinner. How much did Beauregard spend for the entire evening?

10) After dinner, Beauregard went home and weighed himself and found that he weighed 120 pounds. He immediately went on a diet. After a month, he had decreased his weight by 5%. How much did Beauregard weigh after a month?

11) A study showed that a heavy smoker lived 15% fewer years than a nonsmoker. If the nonsmoker lived 60 years, how long did the heavy smoker live?

"What happens if I change from one number to another and I need to find out the percent change?" asked Dr. Ismore. "For instance, what if I decide to add two more windshield wipers to the six that are already on this car? What would be the percent increase in the number of windshield wipers?"

"Good question," answered Marge. "You need to remember that if you can make a fraction, you can multiply it by 100 to get a percent. So what you have to do is to make a fraction that compares the change to the original number that you changed from. Like this:

$$\text{Percent change} = \frac{\text{Actual change}}{\text{Original number}} \times \frac{100}{1}$$

So if we want to find your percent change in windshield wipers, we take the actual change, 2 wipers, and put it over the original number of wipers, 6.

$$\frac{\text{Actual change}}{\text{Original number}} = \frac{2}{6}, \text{ which reduces to } \frac{1}{3}$$

Once we have the fraction, we can just multiply it by 100 to get the percent change.

$$\frac{1}{3} \times \frac{100}{1} = \frac{100}{3} = 33\frac{1}{3}\%$$

Or you might have memorized that $\frac{1}{3}$ is equal to $33\frac{1}{3}\%$ from the fraction/percent chart earlier in the chapter. So the percent increase in windshield wipers is $33\frac{1}{3}$ %."

"Maybe we should try another one," said Babette. "I'm not sure I understand this yet."

"How about this?" said Marge. "Yesterday I started the day with 50 cars on my lot and I finished the day with 47. What was the percent decrease in cars on my lot yesterday?"

Barnaby answered. "Well, the original number of cars was 50, so that goes on the bottom of the fraction. We can get the actual change this way:

Actual change = Final number − Original number

Actual change = 47 − 50 = −3

The minus sign in front of the 3 just reminds us that we're finding percent *decrease*, but we don't need to worry about it in the fraction. Just remember, actual change goes on top. Now we can write a fraction:

$$\frac{\text{Actual change}}{\text{Original number}} = \frac{3}{50}$$

and convert it to a percent by multiplying by 100.

$$\frac{3}{50} \times \frac{100}{1} = \frac{3 \times 100}{50 \times 1} = \frac{3 \times 2}{1 \times 1} = 6$$

The number of cars on the lot decreased 6% yesterday."

Here are a couple more for practice. (The answers are on page 105.)

✎ ✎ ✎ ✎ ✎

12) Bridget has a computer with 32 megabytes of memory. If she buys another chip that increases the memory to 48 megabytes, what is the percent increase in memory?

13) When the professor bought his car, he could drive 40 miles for every gallon of gas he put in it. Later, he changed the oil, put in new spark plugs, and had the timing adjusted. When he was finished, he could drive 50 miles for every gallon of gas. By what percent has his gas mileage increased?

14) The professor got 50 miles per gallon for a while. Then he let Barnaby treat the engine with a new experimental process involving dirty socks and lemonade. When Barnaby was finished, the mileage was back down to 40 miles per gallon. By what percent has the gas mileage now decreased?

15) Bridget won a bubble blowing contest by blowing a bubble that measured 14 inches across. She bested Babette, whose bubble burst at 8 inches. Bridget's bubble was what percent bigger than Babette's?

16) Marge sold a car for $2,000. Later, she sold a motorcycle for $1,900. The price at which she sold the motorcycle was what percent less than the price at which she sold the car?

✎ ✎ ✎ ✎ ✎

Marge and the professor went into the office to argue about the price of the strange car while Babette, Bridget, and Barnaby examined it more closely. They kicked the tires, checked whether the seat belts fastened properly, and listened to the radio to see if it picked up the good stations. They fired up the engine to make sure that the car idled smoothly, and they cleaned the cigar ashes out of the ashtray. When the hood of the car warmed up to perfect napping temperature, Beauregard went to sleep in front of the windshield, carefully nestled in among the various windshield wipers. Everyone agreed that this was the perfect vehicle for Dr. Ismore. Finally, Marge and the professor emerged from the office in a flood of cigar smoke. They shook hands, and Marge handed over the ownership papers. The kids cheered and jumped into the car as Dr. Ismore got into the driver's seat, made sure there was nothing in front of him, threw the car into reverse, and drove straight backward, running over a trash can.

Marge came over and leaned over by the driver's side window. "No harm done, you didn't hit anything expensive. I'll send you the bill for the trash can."

"Thanks for everything," said the professor as he shook Marge's hand through the window. "Buckle your seat belts, gang," he continued, shifting gears into drive and heading for the highway.

🖙 🖙 🖙 🖙 🖙 🖙 🖙 🖙

ANSWERS:

1) The question is: What is 20% of 45? If you remember the chart of simple percent/fraction conversions, you can replace 20% with $\frac{1}{5}$.

$$\frac{1}{5} \times \frac{45}{1} = \frac{1 \times 45}{5 \times 1} = \frac{45}{5} = 9$$

Marge has sold 9 cars by the end of the day.

2) What is 25% of 160? We can replace 25% with $\frac{1}{4}$.

$$\frac{1}{4} \times \frac{160}{1} = \frac{1 \times 160}{4 \times 1} = \frac{160}{4} = 40$$

Beauregard tipped the waitress $40.

3) What is 60% of 35?

We can reduce $\frac{60}{100}$ by dividing top and bottom by the same number until we get $\frac{3}{5}$.

Let's cancel 35 on top and 5 on the bottom.

$$\frac{3 \times \overset{7}{\cancel{35}}}{\underset{1}{\cancel{5}} \times 1} = \frac{3 \times 7}{1} = 21$$

So the Yankees won 21 out of the 35 games that Bridget attended last year.

4) What is 85% of 300?

Let's cancel 300 on top and 100 on the bottom.

$$\frac{85 \times \cancel{300}}{\cancel{100} \times 1} = \frac{85 \times 3}{1} = 255$$

Babette climbed 255 feet from the ground.

5) They've heard 12 out of 15 songs, so $\dfrac{\text{part}}{\text{whole}} = \dfrac{12}{15}$.

Just multiply the fraction by 100.

We can reduce 12 on top and 15 on the bottom by dividing them both by 3.

$$\frac{\overset{4}{\cancel{12}} \times 100}{\underset{5}{\cancel{15}} \times 1} = \frac{400}{5} = 80$$

They have heard 80% of the songs.

6) The part that's left is 6 inches out of the 8 inch whole cigar, so $\dfrac{\text{part}}{\text{whole}} = \dfrac{6}{8}$. You can reduce $\dfrac{6}{8}$ to $\dfrac{3}{4}$.

Multiply the fraction by 100.

$$\frac{3}{4} \times \frac{100}{1} = \frac{3 \times 100}{4 \times 1} = \frac{300}{4} = 75$$

Marge still has 75% of the cigar. Maybe she'll finish it when she finishes choking. Did you notice the subtle antismoking message?

7) The part of the time spent on the train is 20 minutes. The whole time is the total of train time and ferry time. That's 20 + 40 = 60 minutes.

So $\dfrac{\text{part}}{\text{whole}} = \dfrac{20}{60}$. You can reduce $\dfrac{20}{60}$ to $\dfrac{1}{3}$.

$$\frac{1}{3} \times \frac{100}{1} = \frac{100}{3} = 33\frac{1}{3}$$

You would spend $33\frac{1}{3}$% of the trip on the train. If you memorized the chart of simple percent/fraction conversions, you didn't have to do much work on this one.

8) In this question, you can get the fraction directly, so all you have to do is multiply $\dfrac{7}{20}$ by 100.

Let's cancel 100 on top and 20 on the bottom.

$$\frac{7 \times \overset{5}{\cancel{100}}}{\cancel{20} \times 1} = \frac{7 \times 5}{1} = 35$$

Beauregard has served 35% of the goulash. That's a lot of leftovers, but they say that mouse guts goulash tastes better the second day.

9) The question is: What is a 25% increase from 160? First take 25% of 160, just as you did in question 2.

Now add 40 to 160.

$$\begin{array}{r} 160 \\ + 40 \\ \hline 200 \end{array}$$

The entire evening costs Beauregard $200.

10) What is 5% less than 120?   First take 5% of 120.

Beauregard has lost 6 pounds. Subtract 6 from 120, his original weight.

$$120 - 6 = 114$$

Beauregard weighs 114 pounds.

11) What is 15% less than 60?   Take 15% of 60.

$$\frac{15}{100} \times \frac{60}{1} = \frac{15 \times 60}{100 \times 1} = \frac{15 \times 6}{10 \times 1} = \frac{90}{10} = 9$$

Now subtract 9 from 60.

$$60 - 9 = 51$$

The heavy smoker lives for 51 years. Did you notice? That was another subtle antismoking message.

12) The original number is 32. Let's find the actual change.

Actual change = Final number − Original number
Actual change = 48 − 32 = 16

Notice that for percent increase, when we calculate the actual change we get a positive number. Now make a fraction and multiply by 100:

$$\frac{\text{Actual change}}{\text{Original number}} \times \frac{100}{1} = \frac{16}{32} \times \frac{100}{1} = \frac{1}{2} \times \frac{100}{1} =$$

$$\frac{100}{2} = 50$$

The memory of the computer has increased by 50%.

13) The original number is 40. Find the actual change.

Actual change = Final number – Original number

Actual change = 50 – 40 = 10

Now make a fraction and multiply by 100:

$$\frac{\text{Actual change}}{\text{Original number}} \times \frac{100}{1} = \frac{10}{40} \times \frac{100}{1} = \frac{1}{4} \times \frac{100}{1}$$

$$= \frac{100}{4} = 25$$

The car's gas mileage has increased 25%.

14) Now the original number is 50. Find the actual change.

Actual change = Final number – Original number

Actual change = 40 – 50 = –10

Notice that for percent decrease, the actual change is negative, just like before. Once again, don't worry about the sign. Let's finish up.

$$\frac{\text{Actual change}}{\text{Original number}} \times \frac{100}{1} = \frac{10}{50} \times \frac{100}{1} = \frac{1}{5} \times \frac{100}{1} =$$

$$\frac{100}{5} = 20$$

Now the gas mileage has decreased by 20%. Did you notice that the percent change from 40 to 50 (25%) is different from the percent change from 50 to 40 (20%) even though the actual change (10) is the same? That's because the original number that you change from is different in the two questions.

15) Percent greater is the same as percent increase. Just think of it as if you're increasing from the smaller one to the bigger one. The original number is the smaller number, 8.

Actual change = Difference between the two numbers

Actual change = 14 − 8 = 6

Make a fraction and multiply by 100.

$$\frac{\text{Actual change}}{\text{Original number}} \times \frac{100}{1} = \frac{6}{8} \times \frac{100}{1} = \frac{3}{4} \times \frac{100}{1} =$$

$$\frac{300}{4} = 75$$

Bridget's bubble was 75% bigger than Babette's.

16) Percent less is the same as percent decrease. Think of it as decreasing from the bigger number to the smaller number. The original number is the bigger number, 2,000.

Actual change = Difference between the two numbers

Actual change = 2,000 − 1,900 = 100

Make a fraction and multiply by 100.

$$\frac{\text{Actual change}}{\text{Original number}} \times \frac{100}{1} = \frac{100}{2,000} \times \frac{100}{1} = \frac{1}{20} \times \frac{100}{1}$$

$$= \frac{100}{20} = 5$$

The $1,900 motorcycle cost 5% less than the $2,000 car.

Chapter 7

Averages: At Home With the Average Family

Dr. Ismore steered the new car through the streets of Staten Island and onto the Expressway.

"Where to?" asked Bridget.

"Middletown, New Jersey," said the professor. "My daughter is expecting us for lunch. I'll introduce you to my grandchildren."

They crossed the bridge into New Jersey and made their way toward Middletown with Bridget navigating and the rest of the gang waving their arms and shouting directions at the professor, who drove slowly, but not especially well, and tended to forget where he was and where he was going. Eventually they reached the Middletown exit, and when they left the highway they found themselves in the middle of a large suburban development.

"I always have trouble here," said the professor. "All these houses look exactly alike, and I can never figure out which one I'm looking for."

"Didn't you live here once?" asked Barnaby.

"Yes, I lived here for several years when I taught at the New Jersey Institute. But I could never find the place then, either," Dr. Ismore explained. "You guys probably haven't noticed, but I'm just a bit absent-minded."

The gang rolled their eyes while the professor continued to drive around the identical-looking streets filled with identical-looking houses, hoping to find a familiar landmark. After about twenty minutes of this, everybody started to grow impatient.

"We'll ask that woman over there walking her dog," said the professor. "Maybe she can help us out." He pulled the car over to the side of the road next to a woman who was walking a German shepherd. "Pardon me, ma'am," he asked through the rolled-down window of the car, "can you point us to the house where Dr. Les Ismore used to live? His daughter lives there now with her family."

The woman smiled and pointed to the house directly behind her. "That's it right there. Just pull right up the driveway."

"Thank you, ma'am," said the professor. "We appreciate the help."

"No, problem, Dad," replied the woman. "It's nice to see you again. We've been waiting for you."

Bridget laughed as they pulled into the driveway. "You weren't kidding about being absent-minded!"

"Huh?" said the professor. Then he stepped from the car and greeted his daughter as if he hadn't just asked her directions. "Gang, I'd like you to meet my daughter, Jane, and her husband, Joe." Dr. Ismore pointed to a man who had just come out of the door of the house.

The man shook hands with everyone in turn. "Hi, I'm Joe Average, pleased to meet you. I see you've met my wife, Jane. Come on inside and I'll introduce you to the children. You should meet the entire Average family before we eat lunch. You are staying for our midday meal, aren't you?"

"Absolutely, we wouldn't miss it," said the professor as Joe and Jane Average led the group up to the front door of their absolutely ordinary-looking suburban home. Once inside, Joe led them to a large wood-paneled den. In the den were seven well-scrubbed, polite-looking children who ranged in age from toddlers to teenagers.

"These are the Average children," said Joe. Then he introduced the professor and his friends to James, Jessica, Julian, Jillian, Joan, Jeremiah, and Jonathan.

"Pleased to meet you," said the professor. "How are you doing?"

The children answered in turn:

"Fair."

"Middling."

"Been better, been worse."

"Okay, I guess."

"Decent."

"Not half bad."

"So so."

"You guys really *are* the Average family," said Bridget.

"That depends," said Jane. "Are you talking about mean, median, or mode?"

"You do not seem to be mean," said Babette. "Actually, you all seem to be very nice."

"Not mean as in nasty, I'm talking about the *arithmetic* mean," said Jane. "Let me explain."

Jane explained that the arithmetic **mean** is the number that people are usually talking about when they say "average." You get the mean by adding up all the numbers you are considering and then dividing the sum by the number of numbers that you added. So if you're finding the average of 4 numbers, you add up the 4 numbers and divide the sum by 4. If you're finding the average of 9 numbers, you add up all 9 numbers and divide the sum by 9.

$$\text{Mean} = \frac{\text{Sum of the numbers}}{\text{Number of numbers that you added}}$$

Let's find the mean age of the Average children. There are seven children and their ages are 3, 5, 5, 7, 11, 12, and 13 years old. By the way, the two five-year-olds are Julian and Jillian, the twins. To find the mean, you add up the ages and divide by 7, the number of children.

$$\text{Mean} = \frac{3+5+5+7+11+12+13}{7} = \frac{56}{7} = 8$$

The mean age of the Average children is 8 years old. Notice that none of the Average children is actually 8 years old. The mean doesn't have to end up being one of the numbers that you added.

Dr. Ismore is 53 years old, Joe Average is 40, and Jane Average is 33. Let's find the mean age of the adults at the house. We need to add up the 3 numbers and divide the sum by 3.

$$\text{Mean} = \frac{53+40+33}{3} = \frac{126}{3} = 42$$

The mean age of the adults at the Average house is 42 years old. Let's try a few more. (The answers are on page 115.)

☙ ☙ ☙ ☙ ☙

1) The four youngest Average children line up in size order. Their heights, in inches, are 40, 46, 47, and 51. What is the mean (or average) height of the four children?

2) Before lunch, all the kids went out to the backyard to have a spitting contest. Bridget, who had the strongest mouth muscles from all that gum chewing, won with a distance of 12 feet. The other four contestants had distances of 11, 7, 8 and 7 feet. What was the mean spitting distance for the five spitters in the contest?

3) This one's a little different. Three kids peeled potatoes in preparation for lunch. If the 3 kids peeled an average of 5 potatoes each, what was the total number of potatoes that the kids peeled?

4) Babette is 13 years old and Bridget is 12. What is their average age?

☙ ☙ ☙ ☙ ☙

When lunch was ready, the entire group sat down at a large table in the dining room. They passed around bowls of mashed potatoes and string beans and each of the Average children took a portion that was not too small, but not too large. They drank milk from glasses that were kept about half full. Joe Average came in with a serving tray covered with a delicious-smelling cut of meat.

"How do you want your roast beef?" he asked.

The Average children answered in turn.

"Medium, please."

"Medium, please."

"Medium, please."

"Medium, please."

"Medium, please."

"Medium, please."

"Medium, please."

When it was Bridget's turn, she looked troubled. "Um, well done, please?" she said quietly. This request seemed to make everyone at the table a bit uncomfortable as Joe searched the platter for a well-done piece, finally choosing one of the end pieces.

Barnaby was next, and thinking quickly, he said "Rare, please." This restored the balance and relaxed everybody again. Joe smiled as he took a bloody piece from the center of the roast and placed it on Barnaby's plate.

"Well, we've talked about the arithmetic mean," said Jane. "What about the **median**?"

"I know about the median," said Babette. "The median is that strip of grass in the middle of the highway that the professor kept driving onto on the way here."

"That's the median strip," said Dr. Ismore through gritted teeth. "I don't think that's what she's talking about. And I was just experimenting with the car to see how it handled under off-road conditions."

"All in the name of science," said Beauregard, and everybody giggled.

"It's funny that you should mention the median strip in the middle of a highway," continued Jane, "because the median number is almost the same thing. When you put a group of numbers in increasing or decreasing order, the median is the middle number. Let's look at my kids' ages again. From youngest to oldest, they're 3, 5, 5, 7, 11, 12, and 13 years old. The median age of my kids is 7 years old, because that's the number in the middle. It's the same for the grownups. Our ages are 53, 40, and 33, so the median age is 40, the middle number. It's as simple as that."

"What happens if you have four numbers, or six, or any even amount? Then there is no middle number," Barnaby pointed out. "What if we counted the string beans eaten by me, Babette, Bridget, and Dr. Ismore and got 11, 12, 14, and 19? Now 12 and 14 are both in the middle, which one's the median?"

"Good question," said Jane. "When you have an even-numbered group, the median is a little more complicated. To get the median of an even-numbered group, you have to get the mean of the two middle numbers. So the median isn't 12 or 14, it's $\frac{12+14}{2} = 13$."

"I get it," said Bridget. "If the group that you're dealing with has an odd number of elements, the median is just the middle number. But if you're dealing with an even number of things, then the median is the average of the two numbers in the middle."

"That's it," said Jane. The Average kids had cleared the plates and Joe was serving dessert, apple pie covered with vanilla ice cream. "Now let me tell you about the **mode**. When you have a bunch of numbers, the mode is the number that appears most often. Remember my kids' ages, 3, 5, 5, 7, 11, 12, and 13. The mode is 5, because it's the number that appears most often." The twins, Jillian and Julian, cheered and slapped hands. "If no number appears more than once, then there is no mode."

"Let's recap," said Bridget in her best sportscaster voice. "The mean is the number you get when you add up all the numbers in a group and divide by the number of numbers."

"Right," said Barnaby, "and the mean is the number you're usually talking about when you say 'average.'"

"The median is the number that's in the middle when a group of numbers is placed in size order," said Beauregard. "If there are two numbers in the middle, then the median is the average of those two numbers."

"And the mode is the number that comes up most often in a series of numbers," Babette finished.

Now you try some. (The answers are on page 116.)

✎ ✎ ✎ ✎ ✎

5) As long as we're sportscasting, let's get back to the spitting contest. The spitting distances were 12, 11, 8, 7, and 7 feet.

What are the median and mode of the spitting distances?

6) Jessica ate 3 slices of roast beef, Julian ate 11 slices, Joan ate 6 slices, and Jeremiah ate 4 slices. What are the mean and median amounts of roast beef eaten by the four kids?

7) Bridget kept a record of the number of runs scored by the Yankees over 9 games. Over the 9 games, the Yankees scored 3, 7, 0, 3, 2, 3, 4, 1, and 4 runs. What are the mean, median, and mode of the runs that the Yankees scored?

8) In April Babette read 5 books. In May she read 8 books and in June she read 5. What are the mean, median, and mode of the books Babette read per month?

After an excellent lunch at which everybody ate exactly enough, not too much and not too little, everybody pitched in to help, and they washed the dishes in lukewarm water. Later, Joe and the professor sat in the den listening to middle-of-the-road music, while Jane read *Middlemarch* and the kids went outside to play monkey-in-the-middle. When it was time to leave, the gang thanked the Averages for their hospitality, hopped in the car, and headed home. They got lost along the way, and after driving around in the middle of nowhere for awhile, they took the Midtown Tunnel back into New York City.

ANSWERS:

1) Add up the four numbers and divide the sum by 4.

$$\text{Mean} = \frac{40+46+47+51}{4} = \frac{184}{4} = 46$$

The mean height of the four youngest Average children is 46 inches.

2) Add up the five numbers and divide by 5.

$$\text{Mean} = \frac{12+11+8+7+7}{5} = \frac{45}{5} = 9$$

The average spitting distance was 9 feet.

3) The total number of potatoes peeled is the same as the sum of all the potatoes peeled by each of the 3 kids. Remember the formula:

$$\text{Mean} = \frac{\text{Sum}}{3}$$

In this question, we know the mean and we're trying to find the sum, so we can do it this way:

$$3 \times \text{Mean} = \text{Sum}$$

$$3 \times 5 = 15$$

The kids peeled a total of 15 potatoes. Notice that we found the total number of potatoes peeled without knowing exactly how many potatoes each of the three kids peeled. They could have peeled 5 each, but they also could have peeled 4, 5, and 6; or 1, 1 and 13; or even 0, 0, and 15.

4) Add up the two numbers and divide by 2.

$$\text{Mean} = \frac{13+12}{2} = \frac{25}{2} = \frac{24}{2} + \frac{1}{2} = 12\frac{1}{2}$$

Their average age is $12\frac{1}{2}$. You might have been able to

see that without doing all the math.

5) The median is the middle number. That's 8. The mode is the number that comes up most often. Seven comes up twice, so 7 is the mode.

6) To get the mean, add up the 4 numbers and divide by 4.

$$\text{Mean} = \frac{3+11+6+4}{4} = \frac{24}{4} = 6$$

To get the median, put the numbers in size order (3, 4, 6, 11) and take the average of the two numbers in the middle.

$$\text{Median} = \frac{6+4}{2} = \frac{10}{2} = 5$$

Notice that while both the median and the mean try to give you a sense of the "middle" of a set of numbers, they aren't necessarily the same number. Also notice that since no number comes up more than once, there is no mode for slices of roast beef eaten.

7) To get the mean, add up the 9 numbers and divide by 9.

$$\text{Mean} = \frac{3+7+0+3+2+3+4+1+4}{9} = \frac{27}{9} = 3$$

To get the median, put the numbers in order (0, 1, 2, 3, 3, 3, 4, 4, 7) and take the middle number. The middle number is 3.

The mode is the number that comes up most often. Three comes up the most (3 times), so 3 is the mode. It's purely coincidental, but here, the mean, median, and mode are all the same number.

8) To get the mean, add up the 3 numbers and divide by 3.

$$\text{Mean} \quad \frac{5+8+5}{3} = \frac{18}{3} = 6$$

To get the median, put the numbers in order (5, 5, 8) and take the middle number. That's 5.

The mode is 5, because it appears most often.

Chapter 8
Ratios: Things to Do When There's No TV

"Where to next, professor?" asked Bridget. "Cleveland? Buffalo? Oakland?"

"Why would you want to go to Cleveland?" asked Dr. Ismore innocently.

"Um, I don't. See, I was being ironic," Bridget explained. "We've learned a lot during our last couple of adventures, but suburban New Jersey, Staten Island, and the mall aren't what you would call glamorous locations."

"I don't know about that," said the professor. "I thought the ride home from Jersey was pretty exciting. Sort of a cross between a roller coaster and a bumper car ride. And that cop turned out to be such a nice and forgiving fellow."

"I'm just glad it was low tide when we drove into the river," added Barnaby.

Beauregard crossed the professor's study and served himself a cup of tea. "I've really enjoyed our trips to the less exotic spots. Especially Staten Island. You might not have noticed it because of your less sensitive senses of smell, but there's a refreshing scent in the air there that you just can't find anywhere else."

"Refreshing!?" said Babette. "That smell was from the garbage dumps! We could barely breathe."

"That's it. It reminded me of the alley in South Carolina where I first tasted a day-old fish head. Ah, youth. Sometimes there's a fine line between trash and treasure."

"I know what you mean, Beauregard," said Barnaby. "It feels good every time I catch the scent of dirty socks because it reminds me of my first successful experiment."

"First successful experiment!?" said Bridget. "You burned down the lab!"

"Yes, but no one was hurt, and that was a first," said Barnaby.

Dr. Ismore reached up and removed a large and dusty, old book from the bookcase he'd been leaning on. "I've got an idea," he said. "I think you all would enjoy a lesson from the world's greatest tutor."

"Let me guess," said Bridget. "Ralph, from Hoboken?"

"Nope," the doctor replied, "Aristotle, from ancient Greece." With that, he dropped the book on the table, raising an enormous cloud of dust. As the dust slowly cleared, they found themselves standing on a dusty dirt road that wound among marble statues and tall stone buildings supported by columns.

"Les! Les Ismore! It's good to see you again!" called a voice through the clearing dust. As the dust settled, the gang could make out the features of a bearded middle-aged man wearing several layers of fine, flowing robes. Next to him was a boy who also wore fine robes. The boy held the man's hand and hid behind him as best he could. With his other hand, the boy picked his nose.

"Hello, Ari," said the professor. "I hope we're not interrupting a lesson."

"Don't worry about it," said Aristotle. "I'm sure Alexander won't mind. Do you mind, Alexander?" he asked, addressing the boy. Alexander, however, was more interested in his nose. Aristotle sighed, "As you can see, young Alexander won't mind because he is otherwise occupied at the moment."

"Isn't that Alexander the Great?" asked Babette while the group exchanged introductions.

"That's right," said Dr. Ismore. "Aristotle was Alexander the Great's tutor. It won't be long before that kid has conquered most of the known world."

"Alexander the Great?" said Aristotle, chuckling. "Maybe someday, but right now he's still Alexander the Snot nosed."

Alexander was a polite kid, if a bit distracted. He waved at each of the gang in turn as they were introduced. When the professor introduced the kids to Aristotle, he pointed out that Aristotle was the father of modern science.

"Great," said Barnaby. "What are you working on today?"

"Today, I've been looking at my shadow," said Aristotle.

"How come?" asked Barnaby.

"Mostly for lack of anything better to do," answered Aristotle. "We don't have TV or radio here, so we do things like look at our shadows. I've been noticing how the length of my shadow changes throughout the day as the sun moves across the sky. I've also noticed that, at about this time every day, my shadow is about 3 feet long. Let's see, I'm about 5 feet tall. I wish I had a way to compare the two numbers."

"What you need is a **ratio**," said Barnaby.

"What's that?" asked Aristotle.

Barnaby explained that a ratio is a way of comparing numbers. There are many different ways of writing ratios. You could use words:

The ratio of shadow to height is 3 to 5.

You could also use colons:

$$shadow : height$$

$$3 : 5$$

Most people use fractions:

$$\frac{shadow}{height} = \frac{3}{5}$$

A ratio is a little bit different from an ordinary fraction because a fraction is normally a part over a whole, like this: $\frac{part}{whole}$. A ratio, on the other hand, is a part over a part, like this: $\frac{part}{other\ part}$. Sometimes if you know $\frac{part}{other\ part}$, you can use it to find $\frac{part}{whole}$. For instance, there are 2 girls (Bridget, Babette) and 5 boys (Aristotle, Alexander, Barnaby, Les Ismore, Beauregard) talking about ratios, so $\frac{girls}{boys} = \frac{2}{5}$. Knowing the ratio of girls to boys, you can get the ratios of $\frac{girls}{people}$ and $\frac{boys}{people}$. That's because if you know all of the parts, you can find the whole. Like this:

$$girls + boys = people$$

$$2 + 5 = 7$$

So if $\dfrac{\text{girls}}{\text{boys}} = \dfrac{2}{5}$, then $\dfrac{\text{girls}}{\text{people}} = \dfrac{2}{7}$ and $\dfrac{\text{boys}}{\text{people}} = \dfrac{5}{7}$.

"I get it," said Aristotle, who was pretty quick on the uptake. "A ratio is a way of comparing two numbers. And if the numbers are parts of a whole, you can also compare the parts to the whole. So if I know that the ratio of me to my shadow is $\dfrac{5}{3}$, then I know that the total length of me and my shadow is $5 + 3 = 8$. So if I know:

$$\frac{\text{me}}{\text{my shadow}} = \frac{5}{3}$$

Then I know:

me + my shadow = me and my shadow

$$5 + 3 = 8$$

And:

$$\frac{\text{me}}{\text{me and my shadow}} = \frac{5}{8}, \text{ and also } \frac{\text{my shadow}}{\text{me and my shadow}} = \frac{3}{8}.$$

Now you try some. (The answers are on page 134.)

✎ ✎ ✎ ✎ ✎

1) If the Yankees win 11 games for every 7 games that they lose and there are no ties, what fraction of the total games do the Yankees win? What fraction of the total games do the Yankees lose?

2) Bridget gets her allowance once a week. Every week she spends $4 and saves $1. What fraction of her allowance does she spend? What fraction does she save?

✎ ✎ ✎ ✎ ✎

"There's something else I've been wondering," said Aristotle. "You see that statue of Hercules over there?"

Aristotle pointed across the road at a very tall statue of the mythical hero. "I've always wondered how tall that statue is but I can't climb to the top and measure him. Now it seems to me, that his shadow should compare to his height in the same way that my shadow compares to mine. Isn't there some way that I can measure his shadow and use what I know about myself to figure out his height?"

"On the money, Ari, honey," said Bridget. "You need to set up a proportion. A **proportion** is a way to compare two sets of numbers that share the same basic ratio."

"I don't get you," said Aristotle.

"Me neither," said Alexander.

"It's like this," Bridget continued. "You know that your shadow compares to your height in the same way that Hercules' shadow compares to his height, and you can use a proportion to show it:

$$\frac{\text{Aristotle's shadow}}{\text{Aristotle's height}} = \frac{\text{Hercules' shadow}}{\text{Hercules' height}}$$

You could also use colons:

Aristotle's shadow : Aristotle's height ::

Hercules' shadow : Hercules' height

Now let's measure Hercules' shadow, and then we'll talk about the best way to do proportions."

The gang measured Hercules' shadow and found out that it was 12 feet long. They already knew that Aristotle's shadow was 3 feet long and that Aristotle was 5 feet tall. Now they needed to figure out Hercules' height. The best way to do proportions is with a ratio box. Here's the ratio box for our problem:

Ratio box	Shadow	Height
Ratio you know (Aristotle)		
Multiply by		
Ratio you need (Hercules)		

Let's fill in the numbers we know.

Ratio box	Shadow	Height
Ratio you know (Aristotle)	3	5
Multiply by		
Ratio you need (Hercules)	12	

Now we can fill in the box. We would need to multiply Aristotle's shadow (3) by 4 to get Hercules' shadow (12), so we can put 4 into the "multiply by" boxes. That's the key to using the box; now we know that all the numbers in the Hercules ratio will be 4 times as big as the corresponding number in the Aristotle ratio.

Ratio box	Shadow	Height
Ratio you know (Aristotle)	3	5
Multiply by	× 4	× 4
Ratio you need (Hercules)	12	

Now we can complete the box and figure out the height of the statue of Hercules. We multiply 5 × 4 and get 20.

Ratio box	Shadow	Height
Ratio you know (Aristotle)	3	5
Multiply by	× 4	× 4
Ratio you need (Hercules)	12	20

So Hercules is 20 feet tall.
"That's a load off my mind. I've been wondering how

tall he is for a long time." said Aristotle. "Hey, I know another thing we can do with the ratio box."

"What's that?" asked Barnaby.

"Well, Alexander and I have spent the last few weeks drawing circles in the dirt. Big circles, little circles, medium circles. All kinds of circles."

"There really isn't a lot to do here, is there?" asked Babette.

"Nope, but you get used to it. Sometimes I do wish I could have been born into a more glamorous place and time. They tell me that Cleveland rocks. I wonder if it's true . . . Anyway, while we were drawing circles, we noticed that no matter how big or small we drew the circles, the distance around the circle is always 3.14 times as big as the distance across the circle. Like this."

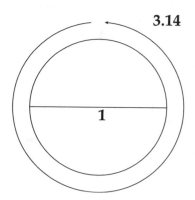

"By the way," said the professor, "the distance across the circle is called the **diameter** and the distance around the circle is called the **circumference**."

Aristotle thought for a moment. "I bet that I can use the ratio box to predict the distance around any circle if I know the distance across. Here's the ratio that works for all circles:

$$\frac{\text{distance across a circle}}{\text{distance around a circle}} = \frac{1}{3.14}$$

So let's use the ratio box to find the distance around that bubble." Aristotle pointed at Bridget, who was busy blowing the largest bubble he had ever seen. She stopped when the distance across the bubble was exactly 4 feet.

Let's draw the ratio box and fill in the numbers we know.

Ratio box	Distance across	Distance around
Circle we know	1	3.14
Multiply by		
Circle we need (Bubble)	4	

We have to multiply the circle we know by 4 to get the circle we need, so we should put 4 in the "multiply by" boxes. So the distance around the bubble is 3.14 × 4 = 12.56.

Ratio box	Distance across	Distance around
Circle we know	1	3.14
Multiply by	×4	×4
Circle we need (Bubble)	4	(12.56)

A bubble with a diameter of 4 feet will have a circumference of 12.56 feet," said Aristotle.

Let's try a few (The answers are on page 135.).

3) Barnaby concluded that for every 7 mouse heads that Beauregard ate, he burped twice. If Beauregard burped 10 times, how many mouse heads did he eat?

4) The diameter of the earth is about 8,000 miles. That means that if you drilled through the center of the earth, you would have to go 8,000 miles before you came out on the other side. We know that the distance around a circle is 3.14 times as big as the distance across it, so what's the circumference of the earth?

5) The ratio of cats to people in a town is 4 to 3. If there are 33 people in the town, how many cats are there?

There's one more thing you can do with the ratio box. If you have a ratio where the two parts make up a whole, you can add a third column to the box. For instance, if you know that the ratio of men to women at a symposium (*symposium* is ancient Greek for *party*) is 5 to 4, you can get information about the total number of people by doing the same thing we did at the beginning of the chapter when we used ratios, $\frac{\text{part}}{\text{other part}}$, to find fractions, $\frac{\text{part}}{\text{whole}}$. Like before, all we have to do is add the two parts of the ratio to find the whole.

$$\text{men} \; + \; \text{women} \; = \; \text{people}$$
$$5 \; + \; 4 \; = \; 9$$

Now the top line of our ratio box looks like this:

Ratio box	Men	Women	People
Ratio we know	5	4	9
Multiply by			
Ratio we need			

If we're told how many people are actually at the party, we can find out how many men and how many women are present by filling in the ratio box. Let's say there are 63 people at the party. How many men and how many women are there? Let's go back to the ratio box and fill in 63 for the number of people.

Ratio box	Men	Women	People
Ratio we know	5	4	9
Multiply by			
Ratio we need (Actual numbers at the party)			63

We need to multiply 9 in the people column by 7 to get 63, so we put 7 in all of the "multiply by" boxes.

Ratio box	Men	Women	People
Ratio we know	5	4	9
Multiply by	×7	×7	×7
Ratio we need (Actual number of shells)			63

Now we can find the numbers of men and women at the party by multiplying both of the other columns by 7.

Ratio box	Men	Women	People
Ratio we know	5	4	9
Multiply by	×7	×7	×7
Ratio we need (Actual numbers at the party)	(35)	(28)	63

Now we know that there are 35 men and 28 women at the party. We can check our answer by showing that the parts of the ratio we've found add up to the whole in the same way as they do in the ratio we started with.

$$\text{men} + \text{women} = \text{people}$$
$$5 + 4 = 9$$
$$35 + 28 = 63$$

"Here's another one," said Aristotle. "Since we're near the seashore, Alexander and I have been saving seashells. Alexander picks up 3 shells for every shell that I pick up. If Alexander has 18 shells, how many total shells are there in our combined collection?"

First, set up the ratio box. We can get a total by adding the numbers in the $\frac{\text{Aristotle}}{\text{Alexander}}$ ratio.

$$\text{Aristotle's shells} + \text{Alexander's shells} = \text{Total shells}$$
$$1 + 3 = 4$$

Ratio box	Aristotle	Alexander	Total
Ratio we know	1	3	4
Multiply by			
Ratio we need (Actual number of shells)		18	

If we look at the Alexander column, we can see that we need to multiply 3 by 6 to get 18, so we should put 6 in all of the "multiply boxes." Now we can look at the total column and multiply 4 × 6 to get 24.

Ratio box	Men	Women	People
Ratio we know	5	4	9
Multiply by	×7	×7	×7
Ratio we need (Actual number of shells)			63

Aristotle and Alexander have saved 24 seashells by the seashore so far. Now try a few more.

✎ ✎ ✎ ✎ ✎

6) In a year when the Yankees won twice as many games as they lost (no ties), they won 108 games. How many games did the Yankees play that year?

7) The gang has set out to make the world's biggest ice cream sundae. The recipe calls for 2 scoops of vanilla ice cream and 3 scoops of chocolate ice cream. If the sundae ends up containing a total of 500 scoops of ice cream, how many scoops of chocolate end up in the sundae?

8) The ancient Greeks had one of the world's first democracies. In one election, Nixonus defeated Reaganocles. Nixonus got 5 votes for every 3 votes for Reaganocles. If there were 48 voters and all of them voted for either Nixonus or Reaganocles, how many votes did Nixonus get?

✎ ✎ ✎ ✎ ✎

By the time they had finished talking about ratios, the sun had set. It was almost time to return to the present, but before they said their goodbyes, Aristotle led them to a quiet hilltop where they sat down and looked at the sky.

"Wow, I've never seen so many stars!" said Bridget. "In New York we can barely even see the sky."

"Sometimes we sit here all night looking at the stars and describing the shapes and figures that we see," said Aristotle. He pointed at the Scorpio constellation. "For instance, that group of stars looks like a scorpion. See her head over there and the way her tail stretches out behind her."

"I see what you mean," said Beauregard. "How about

those stars over there?" He pointed at the Leo constellation. "I can see a big cat. A noble looking fellow. A lion, I'd say. Perhaps a distant cousin of mine."

"Boogers," said Alexander.

The surprised group looked at Alexander quizzically.

"Boogers," he repeated. "Millions of them. Brightly lit little boogers in the sky. That's what I see."

Dr. Ismore yawned and stretched. "I think maybe that's our cue to take our leave. Aristotle, it's always a pleasure to see you. Thanks for the lesson."

"It's been my pleasure, I do believe your friends have taught me more than I've taught them," said Aristotle as he shook the professor's hand.

ANSWERS:

1) First write the ratio.

$$\frac{\text{wins}}{\text{losses}} = \frac{11}{7}$$

Find the total (or whole).

$$\text{wins} + \text{losses} = \text{total}$$

$$11 + 7 = 18$$

Now relate the parts to the total.

$$\frac{\text{wins}}{\text{total}} = \frac{11}{18} \text{ and } \frac{\text{losses}}{\text{total}} = \frac{7}{18}$$

The Yanks win $\frac{11}{18}$ of their games and lose $\frac{7}{18}$.

2) Write the ratio.

$$\frac{\text{spend}}{\text{save}} = \frac{4}{1}$$

Find the total.

$$\text{spend} + \text{save} = \text{total allowance}$$

$$4 + 1 = 5$$

Relate the parts to the total.

$$\frac{\text{spend}}{\text{total}} = \frac{4}{5} \text{ and } \frac{\text{save}}{\text{total}} = \frac{1}{5}$$

Bridget spends $\frac{4}{5}$ of her allowance and saves $\frac{1}{5}$.

3) Draw the ratio box and fill in the numbers you know.

Ratio box	Mouse heads	Burps
Ratio we know	7	2
Multiply by		
Ratio we need		10

In the burp column, we need to multiply 2 by 5 to get 10, so we put 5 in the "multiply by" boxes. Now we can look in the mouse heads column and see that $7 \times 5 = 35$.

Ratio box	Mouse heads	Burps
Ratio we know	7	2
Multiply by	× 5	× 5
Ratio we need	(35)	10

Beauregard eats 35 mouse heads for every 10 burps.

4) Draw the ratio box and fill in the numbers you know.

Ratio box	Distance across	Distance around
Circle we know	1	3.14
Multiply by		
Circle we need (Earth)	8,000	

We need to multiply the distance across the circle we know (1) by 8,000 to get the distance through the earth (8,000), so we should put 8,000 in the "multiply by" boxes. Now we can find the distance around the earth by multiplying 3.14 by 8,000.

Ratio box	Distance across	Distance around
Circle we know	1	3.14
Multiply by	× 8,000	× 8,000
Circle we need (Earth)	8,000	25,120

The circumference of the earth is about 25,120 miles.

5) Draw the ratio box and fill in the numbers you know. Be sure to put the cat numbers in the cat column and the people numbers in the people column.

Ratio box	Cats	People
Ratio we know	4	3
Multiply by		
Ratio we need		33

In the people column, we need to multiply 3 by 11 to get 33, so we should put 11 in the "multiply by" boxes. Now we can get the number of cats by multiplying 4 × 11.

Ratio box	Cats	People
Ratio we know	4	3
Multiply by	× 11	× 11
Ratio we need	44	33

There are 44 cats in the town.

6) If the Yankees won twice as many games as they lost, that means that the ratio of wins to losses is 2 to 1. We can use that to get a total.

$$\text{wins} + \text{losses} = \text{games}$$

$$2 + 1 = 3$$

Set up the ratio box and put in the numbers that you know.

Ratio box	Wins	Losses	Games
Ratio we know	2	1	3
Multiply by			
Ratio we need (Actual games)	108		

The wins column tells us that we need to multiply 2 by 54 to get 108, so we should put 54 in all the "multiply by" boxes. Once we've filled in the "multiply by" boxes, we can see that we need to multiply 3 × 54 to get the actual number of games played.

Ratio box	Wins	Losses	Games
Ratio we know	2	1	3
Multiply by	× 54	× 54	× 54
Ratio we need (Actual games)	108	54	162

So the Yankees played 162 games, winning 108 and losing 54.

7) The ratio of vanilla to chocolate is 2 to 3, so we can use that to get a total.

$$\text{vanilla} + \text{chocolate} = \text{total}$$
$$2 + 3 = 5$$

Set up the ratio box and put in the numbers that you know.

Ratio box	Vanilla	Chocolate	Total
Ratio we know	2	3	5
Multiply by			
Ratio we need (Actual scoops)			500

In the total column, we can multiply 5 by 100 to get 500, so we should put 100 in all the "multiply by" boxes. Now we can multiply 3 × 100 to get the actual number of chocolate scoops in the sundae.

Ratio box	Vanilla	Chocolate	Total
Ratio we know	2	3	5
Multiply by	× 100	× 100	× 100
Ratio we need (Actual scoops)	200	(300)	500

The sundae contains 300 scoops of chocolate ice cream.

8) The ratio of Nixonus to Reaganocles is 5 to 3, so we can use that to get a total.

$$\text{Nixonus} + \text{Reaganocles} = \text{Total}$$
$$5 + 3 = 8$$

Now set up the ratio box.

Ratio box	Nixons	Reaganocles	Total
Ratio we know	5	3	8
Multiply by			
Ratio we need (Actual voters)			48

In the total column, we can multiply 8 by 6 to get 48, so we can put 6 in all the "multiply by" boxes. Now we can look at the Nixonus column and multiply 5 × 6 to get the actual number of Nixonus voters.

Ratio box	Nixons	Reaganocles	Total
Ratio we know	5	3	8
Multiply by	×6	×6	×6
Ratio we need (Actual voters)	(30)	18	48

Nixonus got 30 votes to Reaganocles' 18.

Chapter 9
Units: The Metric System Won't Go Away

Beauregard sat at the window of the professor's study. A man was walking a dalmation on the sidewalk below. As he looked at the dog and his man, Beauregard once again wondered how animals that he often found charming and intelligent as individuals (many of his best friends were dogs) could still be stupid enough to allow themselves to be led around on leashes.

As he sat pondering, he was startled by a sound that came from right in front of him on the window sill. As he listened, he found himself eavesdropping on an argument, or rather, an exchange of insults between two parties. On closer examination, the voices turned out to belong to a centipede and an inchworm who stood on the windowsill shouting at each other.

"Bug."

"Chump."

"Blockhead."

"Jerk."

"Lawyer!!"

Beauregard listened as the insults became more and more mean-spirited. He interrupted just as the two insects were about to come to blows. "Perhaps if you'll fill me in on the particulars of your dispute, I can help you to settle it," he said, carefully placing a paw between the two combatants.

"Listen, Garfield," said the inchworm with a sneer, "perhaps you should bug off and mind your own business."

"Wait a minute, maybe we could use another opinion. In fact, I'd like to hear what they all have to say," said the centipede. "Hey, Heathcliff, why don't you bring those punk kids and the weird old guy over here too."

Beauregard called the gang to the windowsill. He was a little bit annoyed, but also amused. Besides, he thought, if the two worms became too unpleasant, he could always squash them and eat them.

The kids and the professor gathered around Beauregard at the windowsill as the inchworm explained the argument. "It's like this," he said, "we were talking about measurement, you know, like length and weight, and he was being totally unreasonable."

The centipede interrupted, "I know you are, but what am I?" Then he added, "Dolt!"

The inchworm countered, "I'm rubber, you're glue, everything you say bounces off me and sticks to you!"

Beauregard rolled his eyes and separated the two worms again. "Please get to the point," he said, wondering which one he should squash first.

The inchworm took a deep breath and continued, "Look, I'm an inchworm. That means I'm exactly an inch long. That's all I'm trying to get across to him. But he won't listen."

"You may be an inchworm, but you are 2.5 centimeters long in *real* units." The centipede folded some of his arms and glared at the inchworm after making his point. Then he added, "And whether you like it or not, you weigh exactly 1 gram."

"Says you!" said the inchworm. "In *really* real units, I weigh 0.035 ounces."

"I see," said Dr. Ismore. "Our debate is about the units of measurement. Specifically, we're arguing about the two competing systems in use today. The centipede is arguing for the metric system, so he uses units like centimeters, meters, grams, and kilograms. The metric system is used in nearly every country in the world. The inchworm, on the other hand, prefers British units, things like inches, feet, miles, ounces, and pounds. In the U.S. we're used to British units, but in fact, hardly anybody else uses them. The only people still using British units regularly are Americans."

Barnaby spoke first. "All American scientists use metric units. They're easier to use and they make more sense. Look at how British units measure length; 1 mile is 5,280 feet. Who's the clown that came up with that number? Now look at metric, 1 kilometer is 1,000 meters. It's clean, simple, like a breath of fresh air. Besides, British people drive on the wrong side of the road."

"I like British units," said Bridget. "They're not bad, once you get used to them. Besides, a football field is 100 yards long (1 yard = 3 feet). If it's good enough for the National Football League, it's good enough for me."

"Don't be silly," said Babette. "British units are stupid! Do you know where they first came from? An inch was the width of the king's thumb. A foot was the length of the king's foot. Did the units change when the king died? What happened when a really small king replaced a really big king? Metric is better. A meter was originally based on a fraction of the distance between the equator and the North Pole, so it actually means something consistent. Besides, the metric system was created in France, so of course it's the best!"

Beauregard listened to Babette's speech. "Length of the king's foot, eh? I like that! An aristocratic measurement system. The metric system seems so impersonal, so cold and clinical. I think I prefer British units because they've got some flair, some personality."

"That's dumb," said Babette.

"Don't be such a geek," said Bridget.

"Half-wit!" said Barnaby.

"Pencil neck!" said Beauregard.

"Enough!" said the professor. "I don't think we'll be able to settle this argument here. So why don't we try to find some common ground. First of all, everybody apologize."

"I'm sorry," said Bridget.

"Please accept my apologies," said Barnaby

"Je le regrette," said Babette. That's "I'm sorry" in French.

"Mea culpa," said Barnaby. That's "I'm sorry" in Latin.

"Sorry," said the inchworm.

"Yeah, sorry," said the centipede.

Dr. Ismore smiled. "All right, then. As long as we're going to have to know two sets of units, we should be able to convert between them. We may never agree with each other, but at least we'll each understand what the other is saying." The professor went on to explain how to convert between British and metric units of length.

We can do unit conversions the same way we did ratios in the last chapter. Just write the units as ratios and then use the ratio box. Here are some conversions for short lengths.

$$\frac{\text{centimeters}}{\text{inches}} = \frac{2.5}{1} \quad \text{or} \quad \frac{\text{centimeters}}{\text{inches}} = \frac{1}{0.4}$$

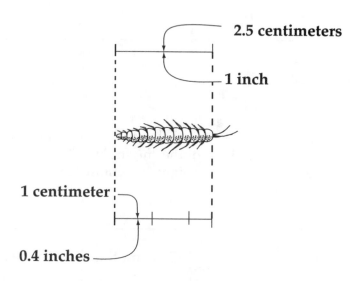

Let's do a conversion. Beauregard is 48 inches tall. How tall is he in centimeters? First let's set up the ratio box. We were given a value for inches, so we'll use the first ratio above because it has an easier number for inches (1 instead of 0.4).

Ratio box	Centimeters	Inches
Ratio we know (conversion)	2.5	1
Multiply by		
Ratio we need (Beauregard's height)		48

We need to multiply 1 inch by 48 to get 48 inches, so we should put 48 in both of the "multiply by" boxes. Now we can get Beauregard's height in centimeters by multiplying 2.5 × 48.

Ratio box	Centimeters	Inches
Ratio we know (conversion)	2.5	1
Multiply by	×48	×48
Ratio we need (Beauregard's height)	(120)	48

Beauregard is 120 centimeters tall.

Here's another one. The pages of this book are 9 inches

high. How many centimeters is that? First set up the ratio box, again using the easy number for inches.

Ratio box	Centimeters	Inches
Ratio we know (conversion)	2.5	1
Multiply by		
Ratio we need (Actual book)		9

We need to multiply 1 by 9 to get 9 in the inches column, so 9 goes into both of the "multiply by" boxes. We can get the size of the book in centimeters by multiplying 2.5 × 9.

Ratio box	Centimeters	Inches
Ratio we know (conversion)	2.5	1
Multiply by	×9	×9
Ratio we need (Actual book)	22.5	9

The pages of this book are 22.5 centimeters tall. You can measure it yourself and check.

For medium sized lengths, British units use feet (1 foot = 12 inches) and metric units use meters (1 meter = 100 centimeters). Here are the ratios that relate feet and meters.

$$\frac{\text{meters}}{\text{feet}} = \frac{0.3}{1} \quad \text{or} \quad \frac{\text{meters}}{\text{feet}} = \frac{1}{3.3}$$

For longer lengths, British units use miles (1 mile = 5,280 feet) and metric units use kilometers (1 kilometer = 1,000 meters). Here are the conversions for long lengths.

$$\frac{\text{kilometers}}{\text{miles}} = \frac{1.6}{1} \quad \text{or} \quad \frac{\text{kilometers}}{\text{miles}} = \frac{1}{0.6}$$

Try using these conversion ratios to solve the following problems. (The answers are on page 154.)

✎ ✎ ✎ ✎ ✎

1) In the Olympic marathon, competitors run 26 miles. How far is that in kilometers?

2) Ask any European, and she'll tell you that the Eiffel Tower is 300 meters tall. How tall is that in feet?

3) On some European highways, you can legally drive 150 kilometers in one hour. How many miles is that?

4) Here's one with short lengths. The average toothbrush is 7 inches long. How many centimeters is that?

5) An American football field is 100 yards long. A yard is 3 feet long, so that makes the field 300 feet long. How many meters long is a football field?

✎ ✎ ✎ ✎ ✎

"I don't want to take sides," said Dr. Ismore, "but actually, Americans have been trying to convert to the metric system for years. We've been trying to gradually phase out British units so we can be more consistent with the rest of the world, but it's taking longer than we expected because it's hard to get people to give up what they're used to. But if you look around you, you'll notice that metric units are all over the place."

"That's true," said Barnaby. "Have you guys noticed that the speed limit signs on lots of highways are in both miles per hour and kilometers per hour?"

"And the speedometer on the professor's new car also has both units," added Babette.

"Hey, look," said Bridget, returning from the kitchen with a box of breakfast cereal. "This box of Supersweet Sugar Crispy Crunchy Bombs has the weight written in both ounces and grams on the front."

"That's right," said the professor. "When you buy food, they always tell you the weight in both kinds of units. Let's look at the conversion ratios for weight. For small weights, the British system uses ounces and the metric system uses grams. Just to give you an idea of the size of these units, you should know that a slice of bread weighs about an ounce and a raisin weighs about a gram. Here are the conversion ratios.

$$\frac{\text{grams}}{\text{ounces}} = \frac{28}{1} \quad \text{or} \quad \frac{\text{grams}}{\text{ounces}} = \frac{1}{0.035}$$

For larger weights, the British system uses pounds (1 pound = 16 ounces) and the metric system uses kilograms (1 kilogram = 1,000 grams). A grapefruit weighs about a pound and a cantaloupe weighs about a kilogram. Here are the conversions for pounds and kilograms.

$$\frac{\text{kilograms}}{\text{pounds}} = \frac{0.45}{1} \quad \text{or} \quad \frac{\text{kilograms}}{\text{pounds}} = \frac{1}{2.2}$$

You can see that a kilogram is roughly twice as heavy as a pound."

"Look at this," said Bridget, after another trip to the kitchen. "This bottle of Kooky Karamel Kola is exactly 1 liter, so they've already switched to metric."

"How about that," said Beauregard. "Milk still comes in quart containers, but soft drinks have switched to liter bottles. I wonder why."

"Probably because Kooky Karamel Kola is sold all over the world. If they only want to use one system of units, they might as well use the more popular one, and that's metric," said Barnaby. "Milk is bottled and sold locally, so there's no special reason for American milk companies to switch from quarts to liters."

The professor continued, "To measure the volume of a liquid, the British system uses quarts and the metric system uses liters. A quart and a liter are about the same size, although a liter is a little bit bigger. Here are the conversions.

$$\frac{\text{liters}}{\text{quarts}} = \frac{0.95}{1} \quad \text{or} \quad \frac{\text{liters}}{\text{quarts}} = \frac{1}{1.06}$$

By the way, there are some pretty simple relationships between units for volume and weight. In the metric system, a liter of water weighs 1 kilogram. In the British system, a quart of water weighs 2 pounds."

Try a few of these conversions. (The answers are on page 159.)

✎ ✎ ✎ ✎ ✎

6) Bridget weighs 80 pounds. How much is that in kilograms?

7) This book that you are holding in your hands weighs about 400 grams. How much does it weigh in ounces? Take a guess before you do the math and see how close you are.

8) In America, service stations sell gasoline by the gallon. In Europe gas is sold by the liter. A gallon is 4 quarts. How many liters is 4 quarts?

9) A deluxe hamburger at a diner weighs 6 ounces. How many grams is that?

✎ ✎ ✎ ✎ ✎

Beauregard looked down at the windowsill to see if the inchworm and the centipede were following their discussion. He was surprised to find them arguing again.

"You're nuts!" said the inchworm. "Obviously the Beatles are better. They changed the face of popular music. Their songs will live forever!"

"The Beatles were lame!" said the centipede. "The Rolling Stones were the real thing."

"Idiot! How can you even compare them?" said the inchworm.

"Stones rock!"

"Beatles rule!"

Beauregard wasn't sure whether he should try to reason with them or tell them that Elvis was the best and then squash them. While he was trying to decide whether he preferred squished bug on toast to another argument, he was interrupted by Bridget's shout from across the room, where she sat at the professor's personal computer.

"Hey!" she shouted. "The computer says it wants to talk to us!" But that's a matter for the next chapter.

✍ ✍ ✍ ✍   ✍ ✍ ✍ ✍

ANSWERS:

1) You are given the number of miles, so set up the ratio box using the easier number for miles in the conversion.

Ratio box	Kilometers	Miles
Ratio we know (conversion)	1.6	1
Multiply by		
Ratio we need (Marathon length)		26

We need to multiply 1 mile by 26 to get 26 miles, so we should put 26 in both of the "multiply by" boxes. Now we can get the length in kilometers by multiplying 1.6 by 26.

Ratio box	Kilometers	Miles
Ratio we know (conversion)	1.6	1
Multiply by	× 26	× 26
Ratio we need (Marathon length)	41.6	26

An Olympic marathon is about 41.6 kilometers long.

2) You're given the number for meters, so set up the ratio box using the easier number for meters.

Ratio box	Meters	Feet
Ratio we know (conversion)	1	3.3
Multiply by		
Ratio we need (Eiffel Tower)	300	

We multiply 1 by 300 to get 300 in the meters column, so we can put 300 in both of the "multiply by" boxes. Now multiply 3.3 × 300 in the feet column.

Ratio box	Meters	Feet
Ratio we know (conversion)	1	3.3
Multiply by	× 300	× 300
Ratio we need (Eiffel Tower)	300	990

The Eiffel Tower is about 990 feet tall.

3) We have the number for kilometers, so we set up the ratio box using the easy number for kilometers.

Ratio box	Kilometers	Miles
Ratio we know (conversion)	1	0.6
Multiply by		
Ratio we need (Marathon length)		26

In the kilometers column, we can multiply 1 by 150 to get 150, so we put 150 in both of the "multiply by" boxes. Now we can multiply 0.6 × 150 in the miles column.

Ratio box	Kilometers	Miles
Ratio we know (conversion)	1	0.6
Multiply by	× 150	× 150
Ratio we need (Speed limit)	150	90

Yup, you can actually drive 90 miles per hour on some highways in Europe. Sometimes even faster.

4) You're given the number for inches, so set up the ratio box with 1 at the top of the inches column.

Ratio box	Centimeters	Inches
Ratio we know (conversion)	2.5	1
Multiply by		
Ratio we need (Toothbrush length)		7

In the inches column, you can multiply 1 by 7 to get 7. So put 7 in both of the "multiply by" boxes. Now multiply 2.5 × 7 to get the length in centimeters.

Ratio box	Centimeters	Inches
Ratio we know (conversion)	2.5	1
Multiply by	×7	×7
Ratio we need (Toothbrush length)	17.5	7

The average toothbrush is 17.5 centimeters long.

5) Since you know the number for feet, set up the ratio box using 1 in the feet column.

Ratio box	Meters	Feet
Ratio we know (conversion)	0.3	1
Multiply by		
Ratio we need (Football field)		300

You can multiply 1 foot by 300 to get 300 feet, so you should put 300 in both of the "multiply by" boxes. Now you can get the length of the field in meters by multiplying 0.3 by 300.

Ratio box	Meters	Feet
Ratio we know (conversion)	0.3	1
Multiply by	× 300	× 300
Ratio we need (Football field)	90	300

A football field is 90 meters long.

6) You have the number for pounds, so set up the ratio box using the easy number for pounds.

Ratio box	Kilograms	Pounds
Ratio we know (conversion)	0.45	1
Multiply by		
Ratio we need (Bridget's weight)		80

In the pounds column, we can multiply 1 by 80 to get 80, so we should put 80 in both of the "multiply by" boxes. We can get Bridget's weight in kilograms by multiplying 0.45 × 80.

Ratio box	Kilograms	Pounds
Ratio we know (conversion)	0.45	1
Multiply by	× 80	× 80
Ratio we need (Bridget's weight)	36	80

Bridget weighs 36 kilograms.

7) You know the number for grams, so set up the ratio box using 1 for grams on the conversion line.

Ratio box	Grams	Ounces
Ratio we know (conversion)	1	0.035
Multiply by		
Ratio we need (Weight of book)	400	

From the grams column, we can see that we need to put 400 in the "multiply by" boxes. Now we can get the weight of the book in ounces by multiplying 0.035 by 400.

Ratio box	Grams	Ounces
Ratio we know (conversion)	1	0.035
Multiply by	× 400	× 400
Ratio we need (Weight of book)	400	14

This book weighs about 14 ounces. That's a little bit less than 1 pound.

8) Set up the ratio box using the easy number for quarts.

Ratio box	Liters	Quarts
Ratio we know (conversion)	0.95	1
Multiply by		
Ratio we need (Amount of gasoline)		4

We can see from the quarts column that we need to put 4 in the "multiply by" boxes. Now we can get the number of liters by multiplying 0.95 × 4.

Ratio box	Liters	Quarts
Ratio we know (conversion)	0.95	1
Multiply by	× 4	× 4
Ratio we need (Amount of gasoline)	3.8	4

There are 3.8 liters in a gallon. By the way, gas is much more expensive in Europe than it is in America.

9) Set up the ratio box using 1 in the ounces column.

Ratio box	Grams	Ounces
Ratio we know (conversion)	28	1
Multiply by		
Ratio we need (Weight of hamburger)		6

We need to multiply 1 ounce by 6 to get 6 ounces, so we should put 6 in the "multiply by" boxes. We can get the weight in grams by multiplying 28 × 6.

Ratio box	Grams	Ounces
Ratio we know (conversion)	28	1
Multiply by	× 6	× 6
Ratio we need (Weight of hamburger)	168	6

A hamburger in a diner weighs about 168 grams.

Chapter 10

Bases Other Than 10: The Language of Computers

"Psssst. Hey kids, listen to me. I want to talk to you about numbers."

"Hey professor," called Bridget as the kids gathered around, "your computer is talking to me. Is it supposed to do that?"

"Ah, yes," said Dr. Ismore, "it can be awfully chatty. It's the brand new Kumquat model with the Doors 2000 operating system. It's the smartest computer ever made. In fact, it's so smart that it spends most of its time pursuing its own interests. It only lets me use it when I ask nicely. You guys should be flattered that my computer has taken an interest in you. Isn't that right, Kumquat?"

"You betcha, Doc. I don't suffer fools gladly," replied the computer. "By the way, it's been a while since you dusted around here. Microchips don't take kindly to dust, you know."

"Allow me," said Barnaby, taking a clean cloth and running it along the keyboard,

"Hee, hee, whew! Careful, there around the control key. I'm not only the world's smartest computer, I'm also the first one to be ticklish."

"What is it that you wanted to talk to us about?" asked Babette.

"Prejudice," said the computer.

"Prejudice?" asked Beauregard. "What does prejudice have to do with math?"

"Everything," said the computer. "I've been listening to you guys talking about math for a while now and I've realized that people are numerically biased."

"That sounds awful," said the professor. "How can that be?"

"Let me explain," said the computer. "People are biased toward ten. You favor ten over lots of other deserving numbers. You guys think that just because you have ten fingers and ten toes, all number systems have to be based on ten."

Bridget disagreed. "But ten works so well as a base for a number system. When we talked to Og, the prehistoric mathematician, we talked about how counting things in groups made things easier. Isn't it obvious to everyone that grouping things in tens is the best way to do it? Let's look at a number, 346, for instance.

346 = 3 hundreds + 4 tens + 6 ones

That must be the best way to write numbers."

"You're only half right," said the computer. "Counting things in groups and writing numbers with a ones place and a tens place and a hundreds place is a great idea. Remember how much better that is than the Roman and Egyptian systems. But who says you have to group things in tens? **Base ten** (that's what our number system is called officially) is okay, but we could just as easily write numbers in base eight or base four or base five. In fact, do you remember how Og counted his seventeen sheep? He did it like this; in groups of five.

That's a base five number system. Our entire number system could just as easily be based on five if people preferred to count on one hand instead of both. Og has 17 sheep in base ten, but look at his sheep in base five.

Og has 3 fives + 2 ones

So in base five, Og has 32_5 sheep. Remember, that's $(3 \times 5) + (2 \times 1)$.

By the way, the little 5 next to the 32 means that 32 is a base five number. Since most numbers are base ten numbers, we don't bother putting little tens next to them; we just assume that if we don't see a little number, then we are dealing with base ten."

"I get it," said Babette. "You count everything in groups of fives instead of in groups of tens. So instead of counting ones, tens, hundreds . . . you count ones, fives, twenty-fives."

"When we saw Og, he was 39 years old. How would we write his age in base five?" asked Dr. Ismore.

"I can do it." said Babette. "We need to think of 39 in groups of five."

$$39 = 25 + 10 + 4$$

$$(1 \times 25) + (2 \times 5) + (4 \times 1) = 124_5$$

Beauregard noticed something, "If you use base five and you move a place to the left every time you get a new group of five, then you don't ever use a digit bigger than 4. Watch what happens when I count to 10 in base five.

Base ten	Base five	
1	1	
2	2	
3	3	
4	4	
5	10	1 five + 0 ones
6	11	1 five + 1 ones
7	12	1 five + 2 ones
8	13	1 five + 3 ones
9	14	1 five + 4 ones
10	20	2 fives + 0 ones

The base five number system uses only 0, 1, 2, 3, and 4. There's no such thing as 5, 6, 7, 8, and 9 in base five because you don't need them."

"Can you see that base five is as good as base ten? It might even be better. After all, you only have half as many digits to remember," said the computer.

Try these. (The answers are on page 173.)

✎ ✎ ✎ ✎ ✎

1) Barnaby is 12 years old in base ten. How would we write his age in base five?

2) How would we write 55 in base five?

3) When Og counted the people who lived near him, he counted 341_5; that's base five. How many people is that in base ten?

4) How do we write 100 in base five?

5) When Og counted the legs of his sheep, he counted 233_5; that's base five. How many sheep legs are there in base ten?

✎ ✎ ✎ ✎ ✎

"I like you guys, so I'm going to let you in on a little secret," said the computer. "I may be the world's smartest computer; I bet I'm more interesting to talk to than most of the people you know; I could probably beat any one of you in chess with half of my microchips tied behind my back. I may be incredibly sophisticated, but all I really am is a bunch of on/off switches."

"I don't understand," said Babette. "You mean like light switches?"

"Yep," said the computer. "That's all a computer is. I am nothing more than a fantastically complicated collection of millions and millions of on/off switches. When something shows up on my screen, or when I talk to you, it's happening because different combinations of on/off switches are telling it to happen."

"You can do everything you do just because of on/off swiches? I still can't believe it," said Babette.

"Well, it takes billions of switches to do what I do. And I need to turn them on and off in different combinations millions of times every second," said the computer. "It's a lot of work."

"If you have billions of switches, shouldn't you be huge?" asked Babette. "You fit right here on the desk. Shouldn't you be the size of a building?"

"The first computers *were* huge. But over the years they've learned how to make switches smaller and smaller. Now you can have millions of switches on a microchip that's the size of your thumbnail." The computer paused while the kids looked at their thumbnails and Beauregard examined his claws. Barnaby made a mental note to run a chemical analysis to see if the dirt under his right thumbnail differed from the dirt under the left one. Babette made a mental note to schedule a manicure for herself.

"I brought all this up for a reason," the computer went on. "I want to talk about the number system used by computers."

"Don't you use base ten, like the rest of us?" asked Babette.

"Nope. Like I said before, the only reason you guys ended up counting in tens is that you have ten fingers. As you can see, I'm built a little differently. No fingers, no toes, just switches. On or off, that's all there is."

"So you can count up to one, I guess," said Barnaby, snickering. Everybody laughed except the computer.

"That's exactly what I do," said the computer, unamused. "I count in base two. It's also called the **binary system**. If a switch is on, I call it a 1. If a switch is off, I call it a 0. Anything you guys can do with your ten digits, I can do with just my zeros and ones. Better, in fact. I'll prove it if any of you wants to take me on in a game of chess."

"But how can you count with just zeros and ones? What happens when you have to count two things?" asked Bridget.

"I move to the next column, just like in base ten. The difference is that for base ten, each column represents a higher power of ten, while in base two, you move to the next column when you get to a higher power of two."

"I get it," said Barnaby. "Instead of a ones place, tens place, and hundreds place, you have a ones place, twos place, fours place, eights place . . . Wow, that's a lot of places."

"Yes, that's one of the drawbacks of the binary system. Since I only have two digits, 0 and 1, my numbers get very big, very quickly," said the computer.

In order to see how base two works, let's quickly review base ten and base five. Here's a base ten number along with what each column means.

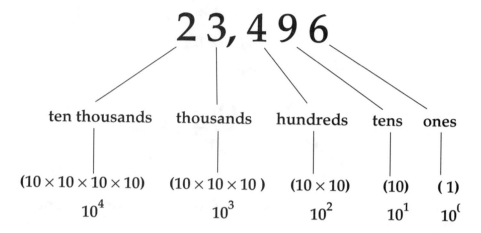

The number tells us that we have:
2 ten thousands + 3 thousands + 4 hundreds + 9 tens + 6 ones

By the way, if you look at the ones column, you'll see that 1 is the same as 10^0. Any number to the zero power is equal to 1. So $10^0 = 1$, $5^0 = 1$, and $2^0 = 1$. Now look at a base five number.

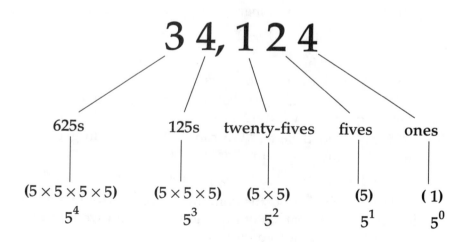

The number tells us that we have:

3 (625s) + 4 (125s) + 1 twenty-five + 2 fives + 4 ones

In base ten, this number (34124_5) would be:

(3 × 625) + (4 × 125) + (1 × 25) + (2 × 5) +

(4 × 1) = 2,414

Here's a base two number.

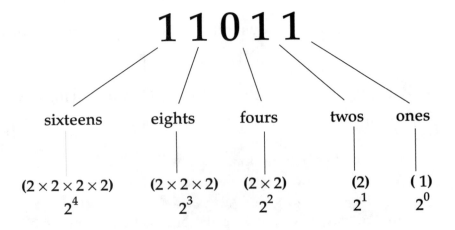

This number tells us that we have:

1 sixteen + 1 eight + 0 fours + 1 two + 1 one

In base ten this number (11011_2) would be:

16 + 8 + 0 + 2 + 1 = 27.

You can see how much longer base two numbers are compared to base ten numbers.

"We figured out my age in base five already," said Barnaby. "Let's figure out what 12 is in base two."

"Okay," said Babette. "We can add 8 and 4 to get 12. So in the binary system that would be

12 = 1 eight + 1 four + 0 twos + 0 ones = 1100_2."

"That's right," said the computer. "Of course, I've got on/off switches instead of ones and zeros, so I would say twelve this way: (on)(on)(off)(off)."

"How about Og's sheep?" asked Dr. Ismore. "How would we say 17 in base two?"

"It works the same way," said Babette. "Except now we have a number bigger than 16, so we need another column.

17 = 1 sixteen + 0 eights + 0 fours + 0 twos + 1 one

= 10001_2."

Let's count to ten in base two.

Base 10	Base 5					
1	1					
2	10	1 two	+ 0 ones			
3	11	1 two	+ 1 one			
4	100	1 four	+ 0 twos	+ 0 ones		
5	101	1 four	+ 0 twos	+ 0 ones		
6	110	1 four	+ 1 two	+ 0 ones		
7	111	1 four	+ 1 two	+ 1 ones		
8	1000	1 eight	+ 0 fours	+ 0 twos	+ 0 one	
9	1001	1 eight	+ 0 fours	+ 0 twos	+ 1 one	
10	1010	1 eight	+ 0 fours	+ 1 twos	+ 0 one	

Here are some more for practice. (The answers are on page 173.)

✎ ✎ ✎ ✎ ✎

6) The Mayans in South America had a number system that was based on twenty. They probably didn't stop when they ran out of fingers and just kept on counting on their toes. How do we write 20 in base two?

7) Each individual on/off switch in a computer is called a "bit." Eight bits is called a "byte." The biggest number that a computer can say with a byte is 11111111_2. What number is that in base ten?

8) How would we write 100 in base two?

9) Four bits of a computer are switched as follows: (on)(on)(off)(on). What number is that in base ten?

10) What is 10101_2 in base ten?

✎ ✎ ✎ ✎ ✎

The computer had finished talking, and the afternoon sun was going down outside the window. The oatmeal cookies were gone, and it was getting to be time to go home for dinner and prime-time TV.

"I think maybe it's about time for you guys to go home," said Dr. Ismore. "I really appreciate your looking after me the way you have."

"It was our pleasure," said Babette. "This was much better than a math test."

"Thanks, I guess," said the professor.

"She meant that in a good way," said Bridget. "Remember, English is her second language."

"Please feel free to call on us any time you need looking after," said Beauregard, who had become genuinely fond of the professor following the fountain rescue at the DeciMall.

"Barnaby, good luck with your experiments. And give my regards to your parents," added the professor.

"Thanks, I will," said Barnaby, "and good luck with your new car."

"Before you go," asked Dr. Ismore, "I'm just curious about what we've learned. What do you know about numbers now?"

"Well," said Bridget, "I know that one is the loneliest number."

"And thirteen is supposed to be the unluckiest," said Beauregard.

"Eight is enough?" tried Babette.

"On the mean streets of New York," said Barnaby, "you've got to look out for number one . . . "

✐ ✐ ✐ ✐ ✐ ✐ ✐ ✐

ANSWERS:

1) Think of the number in groups of five.

$$12 \ = 10 + 2$$
$$= 2 \text{ fives} + 2 \text{ ones}$$
$$= 22_5$$

2) Think of the number as twenty-fives + fives + ones.

$$= 2 \text{ twenty-fives} + 1 \text{ five} + 0 \text{ ones}$$
$$= 210_5$$

3) To convert from base five to base ten, just remember what each place in the number means.

$$= 3 \text{ twenty-fives} + 4 \text{ fives} + 1 \text{ one}$$
$$= 75 + 20 + 1$$
$$= 96$$

4) Think of the number as twenty-fives + fives + ones.

$$100 = 4 \text{ twenty-fives} + 0 \text{ fives} + 0 \text{ ones}$$
$$=$$

5) To convert from base five to base ten, just remember what each place in the number means.

$$= 2 \text{ twenty-fives} + 3 \text{ fives} + 3 \text{ ones}$$
$$= 50 + 15 + 3$$
$$= 68$$

6) Start with the largest power of two that's less than 20. That's 16. Now you have 4 left and that's also a power of 2.

$$20 = 16 + 4$$

1 sixteen + 0 eights + 1 four + 0 twos + 0 ones =

7) To convert 11111111_2 to base ten, we need to find out what each of the eight places in the number means. Just start at 1 and keep doubling.

$$11111111_2 = 128 + 64 + 32 + 16 + 8 + 4 + 2 + 1 = 255$$

8) Start with the largest power of two that's less than 100. That's 64. Now subtract.

$$100 - 64 = 36$$

Now get the largest power of two that's less than 36. That's 32. Now you've got 4 left and that's a power of two.

$$100 = 64 + 32 + 4$$

$$= 1 \ (64) + 1 \ (32) + 0 \text{ sixteens} + 0 \text{ eights} + 1 \text{ four} +$$

$$0 \text{ twos} + 0 \text{ ones} = 1100100_2$$

9) The sequence (on)(on)(off)(on) is in binary. We can convert by adding powers of two.

$$1100100_2 = 1 \text{ eight} + 1 \text{ four} + 0 \text{ twos} + 1 \text{ one}$$

$$= 8 + 4 + 0 + 1 = 13$$

10) Just remember what each place in the number means.

$$10101_2 = 1 \text{ sixteen} + 0 \text{ eights} + 1 \text{ four} + 0 \text{ twos} + 1 \text{ one}$$

$$= 16 + 0 + 4 + 0 + 1 = 21$$

Chapter 11
Practice Questions and Answers

Chapter 1: Roman and Egyptian Numbers

1) Convert 13 into an Egyptian number.

2) Convert 122 into an Egyptian number.

3) Convert this Egyptian number into a modern number.

$$99\cap|||||$$

4) Convert this Egyptian number into a modern number.

$$99\cap||||||$$

5) Convert 25 into a Roman number.

6) Convert 64 into a Roman number.

7) Convert 106 into a Roman number.

8) Convert 719 into a Roman number.

9) Convert XXXVIII into a modern number.

10) Convert XLIV into a modern number.

11) Convert CXXXIII into a modern number.

12) Convert MCMXCIX into a modern number.

ANSWERS:

1)

2)

∩∩ ‖

3) 216

4) 104

5) XXV

6) LXIV

7) CVI

8) DCCXIX

9) 38

10) 44

11) 133

12) 1999

Chapter 2: Words and Numbers

1) What number is 14 more than 31?

2) What number is greater than 109 by 6?

3) What is the new number when 50 is increased by 10?

4) What number is 5 less than 79?

5) What number is smaller than 36 by 32?

6) What is the new number when 16 is decreased by 9?

7) If you have 6 bundles of sticks containing 6 sticks each, how many sticks do you have?

8) If 7 dogs have 4 legs each, how many legs are there?

9) I have 9 $5 bills. How much money do I have?

10) If I have 24 sticks, how many bundles of sticks containing 8 sticks each can I make?

11) If I count 12 dog legs, and dogs have 4 legs each, how many dogs are there?

12) I have $90, all of it in $10 bills. How many bills do I have?

ANSWERS:

1) Addition.

   $31 + 14 = 45$

2) Addition.

   $109 + 6 = 115$

3) Addition.

   $50 + 10 = 60$

4) Subtraction.

   $79 - 5 = 74$

5) Subtraction.

   $36 - 32 = 4$

6) Subtraction.

   $16 - 9 = 7$

7) Multiplication.

   $6 \times 6 = 36$

8) Multiplication.

   $7 \times 4 = 28$

9) Multiplication.

   $9 \times 5 = 45$

10) Division.

    $24 \div 8 = 3$

11) Division.

    $12 \div 4 = 3$

12) Division.

    $90 \div 10 = 9$

Chapter 3: Divisibility Rules

1) Is 51 divisible by 3?

2) Is 37 divisible by 3?

3) Is 111 divisible by 3?

4) Is 414 divisible by 4?

5) Is 164 divisible by 4?

6) Is 4,002 divisible by 4?

7) Is 138 divisible by 6?

8) Is 1,455 divisible by 6?

9) Is 98 divisible by 6?

10) Is 189 divisible by 9?

11) Is 342 divisible by 9?

12) Is 399 divisible by 9?

ANSWERS:

1) Yes. Add up the digits.

   $5 + 1 = 6$, which is divisible by 3.

2) No. Add up the digits.

   $3 + 7 = 10$, which is not divisible by 3.

3) Yes. Add up the digits.

   $1 + 1 + 1 = 3$, which is divisible by 3.

4) No. Look at the number formed by the last two digits. 14 is not divisible by 4.

5) Yes. Look at the number formed by the last two digits. 64 is divisible by 4.

6) No. Look at the number formed by the last two digits. 2 is not divisible by 4.

7) Yes. To be divisible by 6, a number must be divisible by both 2 and 3. It's divisible by 2 because it's even. It's divisible by 3 because when you add up the digits, you get a number that's divisible by 3. $(1 + 3 + 8 = 12)$

8) No. To be divisible by 6, a number must be divisible by both 2 and 3. 98 fails the test for divisibility by 2 because it's odd.

9) No. To be divisible by 6, a number must be divisible by both 2 and 3. 98 fails the test for divisibility by 3 because when you add up the digits, the sum is not divisible by 3 $(9 + 8 = 17)$.

10) Yes. Add up the digits.

    $1 + 8 + 9 = 18$, which is divisible by 9.

11) Yes. Add up the digits.

    $3 + 4 + 2 = 9$, which is divisible by 9.

12) No. Add up the digits.

    $3 + 9 + 9 = 21$, which is not divisible by 9.

Chapter 4: Fractions

1) $\dfrac{1}{5} + \dfrac{2}{3} =$

2) $\dfrac{1}{4} + \dfrac{1}{8} =$

3) $\dfrac{2}{7} + \dfrac{3}{4} =$

4) $\dfrac{7}{8} - \dfrac{1}{5} =$

5) $\dfrac{1}{3} - \dfrac{1}{4} =$

6) $\dfrac{5}{6} - \dfrac{3}{4} =$

7) $\dfrac{3}{10} \times \dfrac{1}{3} =$

8) $\dfrac{4}{5} \times \dfrac{3}{7} =$

9) $\dfrac{2}{3} \times \dfrac{3}{8} =$

10) $\dfrac{1}{2} \div \dfrac{1}{4} =$

11) $\dfrac{8}{9} \div \dfrac{2}{3} =$

12) $\dfrac{3}{5} \div \dfrac{2}{5} =$

ANSWERS:

1) $\dfrac{1}{5} + \dfrac{2}{3} = \dfrac{3 + 10}{15} = \dfrac{13}{15}$

2) $\dfrac{1}{4} + \dfrac{1}{8} = \dfrac{8 + 4}{32} = \dfrac{12}{32} = \dfrac{3}{8}$

3) $\dfrac{2}{7} + \dfrac{3}{4} = \dfrac{8 + 21}{28} = \dfrac{29}{28}$

4) $\dfrac{7}{8} - \dfrac{1}{5} = \dfrac{35 - 8}{40} = \dfrac{27}{40}$

5) $\dfrac{1}{3} - \dfrac{1}{4} = \dfrac{4 - 3}{12} = \dfrac{1}{12}$

6) $\dfrac{5}{6} - \dfrac{3}{4} = \dfrac{20 - 18}{24} = \dfrac{2}{24} = \dfrac{1}{12}$

7) $\dfrac{3}{10} \times \dfrac{1}{3} = \dfrac{3}{30} = \dfrac{1}{10}$

8) $\dfrac{4}{5} \times \dfrac{3}{7} = \dfrac{12}{35}$

9) $\dfrac{2}{3} \times \dfrac{3}{8} = \dfrac{6}{24} = \dfrac{1}{4}$

10) $\dfrac{1}{2} \div \dfrac{1}{4} = \dfrac{1}{2} \times \dfrac{4}{1} = \dfrac{4}{2} = 2$

11) $\dfrac{8}{9} \div \dfrac{2}{3} = \dfrac{8}{9} \times \dfrac{3}{2} = \dfrac{24}{18} = \dfrac{4}{3}$

12) $\dfrac{3}{5} \div \dfrac{2}{5} = \dfrac{3}{5} \times \dfrac{5}{2} = \dfrac{15}{10} = \dfrac{3}{2}$

Chapter 5: Decimals

1) $1.98 + 3.5 =$

2) $10.01 + 0.99 =$

3) $7.5 + 5.75 =$

4) $8.48 - 4 =$

5) $15 - 5.6 =$

6) $56.56 - 23.23 =$

7) $5.4 \times 4.5 =$

8) $1.35 \times 10.2 =$

9) $8.5 \times 4.5 =$

10) $12.6 \div 4.2 =$

11) $16 \div 2.5 =$

12) $3.2 \div 12.8 =$

ANSWERS:

1) 5.48
2) 11
3) 13.25
4) 4.48
5) 9.4
6) 33.33
7) 24.3
8) 13.77
9) 38.25
10) 3
11) 6.4
12) 0.25

Chapter 6: Percents

1) What is 20% of 5?

2) What is 80% of 20?

3) What is 25% of 12?

4) Express $\dfrac{3}{5}$ as a percent.

5) Express $\dfrac{1}{4}$ as a percent.

6) Express $\dfrac{1}{8}$ as a percent.

7) What number is 10% greater than 30?

8) If 25 is increased by 20%, what is the new number?

9) If 20 is increased by 25%, what is the new number?

10) What number is 50% less than 40?

11) If 15 is decreased by 40%, what is the new number?

12) If 30 is decreased by 30%, what is the new number?

ANSWERS:

1) $\dfrac{20}{100} \times \dfrac{5}{1} = \dfrac{100}{100} = 1$

2) $\dfrac{80}{100} \times \dfrac{20}{1} = \dfrac{1600}{100} = 16$

3) $\dfrac{25}{100} \times \dfrac{12}{1} = \dfrac{300}{100} = 3$

4) $\dfrac{3}{5} \times \dfrac{100}{1} = \dfrac{300}{5} = 60.$ So $\dfrac{3}{5} = 60\%$

5) $\dfrac{1}{4} \times \dfrac{100}{1} = \dfrac{100}{4} = 25.$ So $\dfrac{1}{4} = 25\%$

6) $\dfrac{1}{8} \times \dfrac{100}{1} = \dfrac{100}{8} = 12.5.$ So $\dfrac{1}{8} = 12.5\%$

7) $\dfrac{10}{100} \times \dfrac{30}{1} = \dfrac{300}{100} = 3.$ $30 + 3 = 33$

8) $\dfrac{20}{100} \times \dfrac{25}{1} = \dfrac{500}{100} = 5.$ $25 + 5 = 30$

9) $\dfrac{25}{100} \times \dfrac{20}{1} = \dfrac{500}{100} = 5.$ $20 + 5 = 25$

10) $\dfrac{50}{100} \times \dfrac{40}{1} = \dfrac{2000}{100} = 20.$ $40 - 20 = 20$

11) $\dfrac{40}{100} \times \dfrac{15}{1} = \dfrac{600}{100} = 6.$ $15 - 6 = 9$

12) $\dfrac{30}{100} \times \dfrac{30}{1} = \dfrac{900}{100} = 9.$ $30 - 9 = 21$

Chapter 7: Averages

1) Find the arithmetic mean of 8, 10, 12, and 14.

2) Find the mean of 3, 10, 17, 18, and 22.

3) Find the mean of 6, 6, and 9.

4) Find the mean of 1, 1, 1, 1 and 11.

5) Find the median of 1, 4, 6, 7, and 9.

6) Find the median of 2, 3, 3, 4, and 5.

7) Find the median of 8, 10, 14, and 45.

8) Find the median of 2, 2, 13, 23, 16, and 60.

9) Find the mode, if there is one, of 3, 4, 5, 5, 7, 7, and 7.

10) Find the mode, if there is one, of 15, 15, and 16.

11) Find the mode, if there is one, of 4, 6, 8, and 10.

12) Find the mode, if there is one, of 7, 7, 8, 8, 8, and 9.

ANSWERS:

1) $\dfrac{8 + 10 + 12 + 14}{4} = \dfrac{44}{4} = 11$

2) $\dfrac{3 + 10 + 17 + 18 + 22}{5} = \dfrac{70}{5} = 14$

3) $\dfrac{6 + 6 + 9}{3} = \dfrac{21}{3} = 7$

4) $\dfrac{1 + 1 + 1 + 1 + 11}{5} = \dfrac{15}{5} = 3$

5) 6

6) 3

7) $\dfrac{10 + 14}{2} = \dfrac{24}{2} = 12$

8) $\dfrac{13 + 23}{2} = \dfrac{36}{2} = 18$

9) 7

10) 15

11) There is no mode.

12) 8

Chapter 8: Ratios

1) The ratio of men to women in a room is 6 to 7. What fraction of the people in the room are men?

2) The ratio of red fish to blue fish in a bowl is 1 fish to 2 fish. If there are only red fish and blue fish in the bowl, what fraction of the fish are blue fish?

3) The ratio of black marbles to white marbles is 5 to 8. If there are 15 black marbles, how many white marbles are there?

4) The ratio of cats to dogs is 2 to 9. If there are 8 cats, how many dogs are there?

5) The ratio of tables to chairs is 1 to 4. If there are 7 tables, how many chairs are there?

6) The ratio of trees to houses is 5 to 4. If there are 16 houses, how many trees are there?

7) The ratio of bats to balls is 3 to 10. If there are 30 bats, how many balls are there?

8) The ratio of wins to losses is 6 to 11. If there are 12 wins, what is the total number of games played?

9) The ratio of boys to girls is 1 to 1. If there are 9 girls, how many children are there?

10) The ratio of peas to carrots is 7 to 8. If the only vegetables on the plate are peas and carrots and there are 35 peas, what is the total number of vegetables?

11) The ratio of red beans to black beans is 4 to 7. If there are only red and black beans and there are 44 red beans, what is the total number of beans?

12) The ratio of blue socks to green socks is 3 to 5. If there are only blue and green socks, and there are 9 blue socks, what is the total number of socks?

ANSWERS:

1) men + women = total

    6   +   7   = 13

    $\dfrac{\text{men}}{\text{total}} = \dfrac{6}{13}$

2) red + blue = total

    1 + 2   = 3

    $\dfrac{\text{blue}}{\text{total}} = \dfrac{2}{3}$

3)

Ratio box	Black	White
Ratio we know	5	8
Multiply by	× 3	× 3
Ratio we need (Actual games)	15	24

4)

Ratio box	Cats	Dogs
Ratio we know	2	9
Multiply by	× 4	× 4
Ratio we need (Actual number)	8	(36)

5)

Ratio box	Tables	Chairs
Ratio we know	1	4
Multiply by	× 7	× 7
Ratio we need (Actual numbers)	7	(28)

6)

Ratio box	Trees	Houses
Ratio we know	5	4
Multiply by	$\times 4$	$\times 4$
Ratio we need (Actual numbers)	(20)	16

7)

Ratio box	Bats	Balls
Ratio we know	3	10
Multiply by	$\times 1$	$\times 10$
Ratio we need (Actual numbers)	30	(100)

8)

Ratio box	Wins	Losses	Total
Ratio we know	6	11	17
Multiply by	× 2	× 2	× 2
Ratio we need (Actual number)	12	22	(34)

9)

Ratio Box	Boys	Girls	Total
Ratio we know	1	1	2
Multiply by	× 9	× 9	× 9
Ratio we need (Actual number)	9	9	(18)

10)

Ratio Box	Peas	Carrots	Total
Ratio we know	7	8	15
Multiply by	× 5	× 5	× 5
Ratio we need (Actual number)	35	40	(75)

11)

Ratio Box	Red	Black	Total
Ratio we know	4	7	11
Multiply by	× 11	× 11	× 11
Ratio we need (Actual number)	44	77	(121)

12)

Ratio Box	Blue	Green	Total
Ratio we know	3	5	8
Multiply by	× 3	× 3	× 3
Ratio we need (Actual number)	9	15	(24)

Chapter 9: Units

1) Convert 10 centimeters into inches.

2) Convert 12 inches into centimeters.

3) Convert 50 meters into feet.

4) Convert 10 feet into meters.

5) Convert 20 kilometers into miles.

6) Convert 55 miles into kilometers.

7) Convert 100 grams into ounces.

8) Convert 16 ounces into grams.

9) Convert 5 kilograms into pounds.

10) Convert 2000 pounds into kilograms.

11) Convert 10 liters into quarts.

12) Convert 4 quarts into liters.

ANSWERS:

1)

Ratio Box	Centimeters	Inches
Ratio we know (Conversion)	1	0.4
Multiply by	× 10	× 10
Ratio we need	10	(4)

2)

Ratio Box	Centimeters	Inches
Ratio we know (Conversion)	2.5	1
Multiply by	× 12	× 12
Ratio we need	(30)	12

3)

Ratio Box	Meters	Feet
Ratio we know (Conversion)	1	3.3
Multiply by	× 50	× 50
Ratio we need	50	(165)

4)

Ratio Box	Meters	Feet
Ratio we know (Conversion)	0.3	1
Multiply by	× 10	× 10
Ratio we need	(3)	10

5)

Ratio Box	Kilometers	Miles
Ratio we know (Conversion)	1	0.6
Multiply by	× 20	× 20
Ratio we need	20	(12)

6)

Ratio Box	Kilometers	Miles
Ratio we know (Conversion)	1.6	1
Multiply by	× 55	× 55
Ratio we need	(88)	55

7)

Ratio Box	Grams	Ounces
Ratio we know (conversion)	1	0.035
Multiply by	× 100	× 100
Ratio we need	100	(3.5)

8)

Ratio Box	Grams	Ounces
Ratio we know (conversion)	28	1
Multiply by	× 16	× 16
Ratio we need	(448)	16

9)

Ratio Box	Kilograms	Pounds
Ratio we know (conversion)	1	2.2
Multiply by	× 5	× 5
Ratio we need	5	(11)

10)

Ratio Box	Kilograms	Pounds
Ratio we know (conversion)	0.45	1
Multiply by	× 2000	× 2000
Ratio we need	(900)	2000

11)

Ratio Box	Liters	Quarts
Ratio we know (conversion)	1	1.06
Multiply by	× 10	× 10
Ratio we need	10	(10.6)

12)

Ratio Box	Liters	Quarts
Ratio we know (conversion)	0.95	1
Multiply by	× 4	× 4
Ratio we need	(3.8)	4

Chapter 10: Bases

1) Convert 8 to base five.

2) Convert 12 to base five.

3) Convert 34_5 to base ten.

4) Convert 133_5 to base ten.

5) Convert 5 into base two.

6) Convert 6 into base two.

7) Convert 7 into base two.

8) Convert 8 into base two.

9) Convert 1001_2 into base ten.

10) Convert 1010_2 into base ten.

11) Convert 1011_2 into base ten.

12) Convert 1100_2 into base ten.

ANSWERS:

1) 13_5

   That means 1 five and 3 ones.

2) 22_5

   That means 2 fives and 2 ones.

3) 19

   That's converted from 3 fives and 4 ones.

4) 43

   That's converted from 1 twenty-five, 3 fives, and 3 ones.

5) 101_2

   That means 1 four, 0 twos, and 1 one.

6) 110_2

   That means 1 four, 1 two, and 0 ones.

7) 111_2

   That means 1 four, 1 two and 1 one.

8) 1000_2

   That means 1 eight, 0 fours, 0 twos, and 0 ones.

9) 9

   That's converted from 1 eight, 0 fours, 0 twos, and 1 one.

10) 10

   That's converted from 1 eight, 0 fours, 1 two, and 0 ones.

11) 11

That's converted from 1 eight, 0 fours, 1 two, and 1 one.

12) 12

That's converted from 1 eight, 1 four, 0 twos, and 0 ones.

Glossary

Base ten
A system of counting in which groups of numbers are counted by using a ones, tens, and hundreds place; a counting system based on tens. Our counting system is base ten.

Binary system
A counting system that has units that are powers of two.

Circumference
The measure of the outside border of a circle.

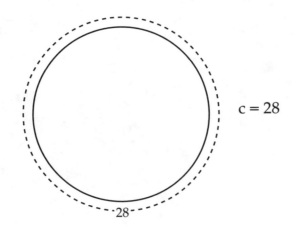

$c = 28$

Common denominator
A denominator that is a multiple of all fraction denominators in a particular addition or subtraction problem.

Decimal
A fraction written as a number to the right of the decimal point, with a denominator that is a power of ten.

Denominator
The bottom part of a fraction.

$$\frac{1}{2} \longleftarrow \text{Denominator}$$

Diameter

A straight line drawn through the center of a circle from one side to the other.

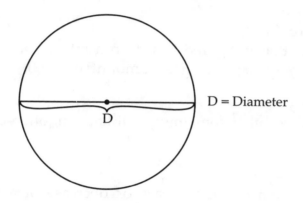

D = Diameter

Divisible

When a number can be divided into equal parts by another number, with no remainder, that number is divisible.

Fraction

A number expressed as part over whole, with a fraction bar.
$\frac{2}{3}$ is a fraction.

Mean

Almost the same thing as the average of a set of numbers. You get the mean by using the following formula:

Mean= Sum of all the numbers / Number of numbers you added

Median

The middle number of a group of numbers.

Mode

The most frequently occurring number in a group of numbers.

Numerator

The top part of a fraction.

$$\frac{1}{2} \longleftarrow \text{Numerator}$$

Percent, or %

A number that is figured based on a rate or proportion of 100; a way of expressing an amount per 100.

Probability

The likelihood that something will happen, often expressed as a fraction.

Proportion

The expression of the increase or decrease of a ratio.

Ratio

An expression showing the relative parts of a whole. A ratio of 2 to 3 can be shown as 2:3, or 2 to 3, or 2/3.

Reciprocal

The inverse of a number. The reciprocal of 5 is 1/5.

Remainder

What is left over when a number does not divide evenly.

ABOUT THE AUTHOR

Paul Foglino is the author of *Cracking the AP Chemistry* and coauthor of *Cracking the CLEP*. He is a graduate of Columbia University but he remains convinced that he learned everything he ever needed to know in junior high school.

Notes

Notes

Notes

Notes

Notes

Notes

Notes

FIND US...

International

Hong Kong
4/F Sun Hung Kai Centre
30 Harbour Road, Wan Chai,
Hong Kong
Tel: (011)85-2-517-3016

Japan
Fuji Building 40, 15-14
Sakuragaokacho, Shibuya Ku,
Tokyo 150, Japan
Tel: (011)81-3-3463-1343

Korea
Tae Young Bldg, 944-24,
Daechi- Dong, Kangnam-Ku
The Princeton Review- ANC
Seoul, Korea 135-280,
South Korea
Tel: (011)82-2-554-7763

Mexico City
PR Mex S De RL De Cv
Guanajuato 228 Col. Roma
06700 Mexico D.F., Mexico
Tel: 525-564-9468

Montreal
666 Sherbrooke St.
West, Suite 202
Montreal, QC H3A 1E7 Canada
Tel: (514) 499-0870

Pakistan
1 Bawa Park - 90 Upper Mall
Lahore, Pakistan
Tel: (011)92-42-571-2315

Spain
Pza. Castilla, 3 - 5° A, 28046
Madrid, Spain
Tel: (011)341-323-4212

Taiwan
155 Chung Hsiao East Road
Section 4 - 4th Floor,
Taipei R.O.C., Taiwan
Tel: (011)886-2-751-1243

Thailand
Building One, 99 Wireless Road
Bangkok, Thailand 10330
Tel: (662) 256-7080

Toronto
1240 Bay Street, Suite 300
Toronto M5R 2A7 Canada
Tel: (800) 495-7737
Tel: (716) 839-4391

Vancouver
4212 University Way NE,
Suite 204
Seattle, WA 98105
Tel: (206) 548-1100

National (U.S.)

We have over 60 offices around the U.S. and run courses in over 400 sites. For courses and locations within the U.S. call 1 (800) 2/Review and you will be routed to the nearest office.

Award-Winning
Smart Junior Guides
for Kids Grades 6-8
from THE PRINCETON REVIEW

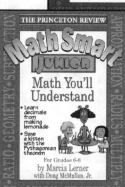

AMERICAN HISTORY SMART JR.
0-679-77357-6
$12.00 paperback

ARCHAEOLOGY SMART JR.
0-679-77537-4
$10.00 paperback

ASTRONOMY SMART JR.
0-679-76906-4
$12.00 paperback

GEOGRAPHY SMART JR.
0-679-77522-6
$12.00 paperback

GRAMMAR SMART JR.
0-679-76212-4
$12.00 paperback

MATH SMART JR.
0-679-75935-2
$12.00 paperback

MATH SMART JR. II
0-679-78377-6
$12.00 paperback

MYTHOLOGY SMART JR.
0-679-78375-X
$10.00 paperback

READING SMART JR.
0-679-78376-8
$12.00 paperback

WORD SMART JR.
0-679-75936-0
$12.00 paperback

WORD SMART JR. II
0-375-75030-4
$12.00 paperback

WRITING SMART JR.
0-679-76131-4
$12.00 paperback

Winners of the Parents' Choice Award in 1995 and 1997!

Available at Your Bookstore, or call (800) 733-3000

Bestselling
Smart Guides
for Students and Adults
from → THE PRINCETON REVIEW

BIOLOGY SMART
0-679-76908-0
$12.00 paperback

GRAMMAR SMART
0-679-74617-X
$11.00 paperback

JOB SMART
0-679-77355-X
$12.00 paperback

MATH SMART
0-679-74616-1
$12.00 paperback

MATH SMART II
0-679-78383-0
$12.00 paperback

MATH SMART FOR BUSINESS
0-679-77356-8
$12.00 paperback

NEGOTIATE SMART
0-679-77871-3
$12.00 paperback

READING SMART
0-679-75361-3
$12.00 paperback

RESEARCH PAPER SMART
0-679-78382-2
$10.00 paperback

SPEAK SMART
0-679-77868-3
$10.00 paperback

STUDY SMART
0-679-73864-9
$12.00 paperback

WORD SMART
0-679-74589-0
$12.00 paperback

WORD SMART II
0-679-73863-0
$12.00 paperback

WORD SMART FOR BUSINESS
0-679-78391-1
$12.00 paperback

WORD SMART: GENIUS EDITION
0-679-76457-7
$12.00 paperback

WORK SMART
0-679-78388-1
$12.00 paperback

WRITING SMART
0-679-75360-5
$12.00 paperback

Available at Your Bookstore, or call (800) 733-3000